# Solid State Lasers Materials, Technologies and Applications

Special Issue Editor
**Federico Pirzio**

MDPI • Basel • Beijing • Wuhan • Barcelona • Belgrade

**MDPI**

*Special Issue Editor*
Federico Pirzio
University of Pavia
Italy

*Editorial Office*
MDPI AG
St. Alban-Anlage 66
Basel, Switzerland

This edition is a reprint of the Special Issue published online in the open access journal *Applied Sciences* (ISSN 2076-3417) from 2017–2018 (available at: http://www.mdpi.com/journal/applsci/special_issues/solid_state_lasers_materials).

For citation purposes, cite each article independently as indicated on the article page online and as indicated below:

Lastname, F.M.; Lastname, F.M. Article title. *Journal Name* **Year,** *Article number*, page range.

**First Editon 2018**

**ISBN 978-3-03842-841-1 (Pbk)**
**ISBN 978-3-03842-842-8 (PDF)**

Cover photo courtesy of Federico Pitzio

# Table of Contents

# About the Special Issue Editor

**Federico Pirzio**, Assistant Professor, received his Ph.D. from the University of Pavia (Italy) in 2007 and was a postdoctoral researcher from 2007 to 2011. In 2011, he joined the Department of Electrical, Biomedical and Computer Engineering at the University of Pavia as an assistant professor. Dr. Pirzio is a member of the research staff of the Laser Source Laboratory. His research topics are the design and numerical modelling of diode-pumped solid-state lasers sources and laser systems in every functioning regime (continuous wave, nanosecond and sub-nanosecond Q-Switching regime, and picosecond and femtosecond mode-locked regime), the design of nonlinear frequency-conversion stages based on harmonic generation, parametric down-conversion, and stimulated Raman scattering, and the characterization of new laser materials and saturable absorbers for ultrafast lasers. His teaching activity includes courses in classical mechanics, classical electromagnetism, and quantum electronics for engineering students at the University of Pavia.

*applied*
*sciences*

MDPI

*Editorial*

# Special Issue on Solid State Lasers Materials, Technologies and Applications

Federico Pirzio

Dipartimento di Ingegneria Industriale e dell'Informazione, Università di Pavia, Via Ferrata 5, 27100 Pavia, Italy; federico.pirzio@unipv.it

Received: 13 March 2018; Accepted: 15 March 2018; Published: 17 March 2018

## 1. Introduction

Even though more than half a century has already passed since the first demonstration of laser action in ruby crystal, solid-state lasers are still a hot research topic. Their unique versatility has made them irreplaceable tools in many fields. The list of actual or potential applications is continuously being updated at a pace that is set by the rate of advancing research. This virtuous cycle is only made possible by the favorable interplay between continuous research progress, enabling new exciting applications, and the emergence of new needs from industrial markets and fundamental research fields, fostering the study of new materials, the development of new technologies, and the fast maturation of the existent ones.

## 2. Solid State Lasers Materials, Technologies and Applications

In light of the above, this special issue was introduced to collect the latest research on relevant topics and reviews on the most recent technology developments and applications. There were 20 papers submitted to this special issue, and 14 papers were accepted. Of these, three were review papers and the remaining 11 were research articles.

Regarding solid-state lasers' materials and technologies, the first paper, authored by N. G. Boetti, D. Pugliese, E. Ceci-Ginistrelli, J. Lousteau, D. Janner and D. Milanese presents a comprehensive review on the recent advances in phosphate fiber laser technology [1]. The second paper provides another review on state of the art high-power solid-state deep ultra-violet lasers, authored by H. Xuan, H. Igarashi, S. Ito, C. Qu, Z. Zhao and Y. Kobayashi [2]. Nonlinear optics applications of high power lasers are presented in [3], where H. Su, Y. Peng, J. Chen, Y. Li, P. Wang, and Y. Leng present a Master Oscillator Power Amplifier (MOPA) laser at 1064 nm, delivering hundreds of mJs, 50-ps pulses at 100 Hz, designed to work as a pump for Optical Parametric Amplification (OPA) experiments. Two other articles describe MOPA laser systems operating at 1 $\mu$m. In particular, a very flexible, temporally programmable, arbitrary pulse shape hybrid MOPA laser was described by M. Nie, Q. Liu, X. Cao, and X. Fu in [4], whereas an industrial grade, picosecond regenerative amplifier based on Nd:YVO$_4$ is presented by Z. Bai, Z. Bai, Z. Kang, F. Lian, W. Lin, and Z. Fan in [5].

Mid-Infrared laser sources are present in two papers of this special issue. In particular, direct nanosecond pulse generation in a Q-Switched Er:Y$_2$O$_3$ ceramic laser is reported by X. Ren, Y. Wang, J. Zhang, D. Tang, and D. Shen in [6], whereas parametric down-conversion is exploited in a cw-OPO based on a periodically-poled Lithium Niobate crystal to generate a multi-Watt output tunable between 2.4–2.9 $\mu$m or 3.14–3.45 $\mu$m, depending on the nonlinear crystal poling period (Y. Liu, X. Xie, J. Ning, X. Lv, G. Zhao, Z. Xie, and S. Zhuin in [7]).

Coherent beam addition is addressed in two papers. D. Malka, E. Cohen, and Z. Zalevsky propose a novel concept for power beam combining in Photonic Crystal Fibers [8], whereas a fast frequency acquisition and phase locking method for high spectral purity nonplanar ring oscillator is presented by Y. Wang, C. Wang, Y. Tao, Y. Liu, Q. Zouh, J. Su, Z. Wang, S. Shi, and Q. Qiu in [9]. The proposed technique can be used to enhance the performance of optical phase locking loops used in space coherent optical communication systems and active coherent laser beam combiners.

A detailed study on the impact of subsurface impurity defects introduced in the polishing process of fused-silica optics, and the beneficial effect of HF acid etching in the defect removal and laser-induced damage threshold enhancement, is presented by J. Cheng, J. Wang, J. Hou, H. Wang, and L. Zhang in [10].

Applications of solid state lasers, in particular in the field of laser welding and laser cladding, are presented in four papers published in this special issue. The first one is a review paper from M. Jiang, W. Tao, and Y. Chen [11]. In this paper, a comprehensive overview on laser welding under vacuum is presented, which is a recently re-discovered laser welding technique [12] that can help in fully exploiting the potential of recently available high-power and high-brightness solid-state fiber and disk lasers. A study on high-power (10 kW) fiber laser welding of stainless steel T-joint focused on the investigation of the impact best fiber core diameter selection is presented by A. Unt, I. Poutiainen, S. Grunenwald, M. Sokolov, and A. Salminen in [13]. The effect of welding position on the quality of the laser welds is investigated in detail by B. Chang, Z. Yuan, H. Pu, H. Li, H. Cheng, D. Du, and J. Shan in [14], where a high-power (6 kW) fiber laser is used to weld titanium alloys. Finally, a study on the effect of Molybdenum content on the microstructures and properties of stainless steel coatings by laser cladding is reported by K. Wang, B, Chang, J, Chen, H. Fu, Y. Lin, and Y. Lei in [15].

## 3. Future Perspectives

The large diversification of the topics covered by the review paper and research articles published in this special issue is clear evidence of the exceptional flexibility, potential and variety of application of solid-state lasers. Based on what we saw happening in recent decades, the research on new materials, the advances in photonics technologies and the increasing demand for speed, cleanliness and high-precision in industrial processes will contribute to propelling the research in this exciting field in the foreseeable future.

**Acknowledgments:** This issue would not be possible without the contributions of various talented authors, hardworking and professional reviewers, and dedicated editorial team of Applied Sciences. I wish to congratulate all the contributors. I am especially grateful to all the reviewers that helped the authors to improve their papers, with high-quality, constructive and punctual reviews. Finally, I would like to place on record my gratitude to Felicia Zhang and to the kind, efficient and professional editorial team of Applied Sciences.

**Conflicts of Interest:** The authors declare no conflict of interest.

## References

1.	Boetti, N.G.; Pugliese, D.; Ceci-Ginistrelli, E.; Lousteau, J.; Janner, D.; Milanese, D. Highly Doped Phosphate Glass Fibers for Compact Lasers and Amplifiers: A Review. *Appl. Sci.* **2017**, *7*, 1295, doi:10.3390/app7121295.
2.	Xuan, H.; Igarashi, H.; Ito, S.; Qu, C.; Zhao, Z.; Kobayashi, Y. High-power, Solid-State, Deep Ultraviolet Laser Generation. *Appl. Sci.* **2018**, *8*, 233, doi:10.3390/app8020233.
3.	Su, H.; Peng, Y.; Chen, J.; Li, Y.; Wang, P.; Leng, Y. A High-Energy, 100 Hz, Picosecond Laser for OPCPA Pumping. *Appl. Sci.* **2017**, *7*, 997, doi:10.3390/app7100997.
4.	Nie, M.; Liu, Q.; Cao, X.; Fu, X. Temporally Programmable Hybrid MOPA Laser with Arbitrary Pulse Shape and Frequency Doubling. *Appl. Sci.* **2017**, *7*, 892, doi:10.3390/app7090892.
5.	Bai, Z.; Bai, Z.; Kang, Z.; Lian, F.; Lin, W.; Fan, Z. Non-Pulse-Leakage 100-kHz Level, high Beam Quality Industrial Grade Nd:YVO$_4$ Picosecond Amplifier. *Appl. Sci.* **2017**, *7*, 615, doi:10.3390/app7060615.
6.	Ren, X.; Wang, Y.; Zhang, J.; Tang, D.; Shen, D. Short-Pulse-Width Repetitively Q-Switched 2.7-μm Er:Y$_2$O$_3$ Ceramic Laser. *Appl. Sci.* **2017**, *7*, 1201, doi:10.3390/app7111201.
7.	Liu, Y.; Xie, X.; Ning, J.; Lv, X.; Zhao, G.; Xie, Z.; Zhu, S. A High-Power Continuous-Wave Mid-Infrared Optical Parametric Oscillator Module. *Appl. Sci.* **2018**, *8*, 1, doi:10.3390/app8010001.
8.	Malka, D.; Cohen, E.; Zalevsky, Z. Design of 4 × 1 Power Beam Combiner Based on MultiCore Photonic Crystal Fiber. *Appl. Sci.* **2017**, *7*, 695, doi:10.3390/app7070695.

*Appl. Sci.* **2018**, *8*, 460

9. Wang, Y.; Wang, C.; Tao, Y.; Liu, Y.; Zouh, Q.; Su, J.; Wang, Z.; Shi, S.; Qiu, Q. Fast Frequency Acquisition and Phase locking of Nonplanar Ring OScillators. *Appl. Sci.* **2017**, *7*, 1032, doi:10.3390/app7101032.

10. Cheng, J.; Wang, J.; Hou, J.; Wang, H.; Zhang, L. Effect of Polishing-Induced Subsurface Impurity Defects on Laser Damage Resistance of Fused Silica Optics and Their Removal with HF Acid Etching. *Appl. Sci.* **2017**, *7*, 838, doi:10.3390/app7080838.

11. Jiang, M.; Tao, W.; Chen, Y. Laser Welding under Vacuum: A Review. *Appl. Sci.* **2017**, *7*, 909, doi:10.3390/app7090909.

12. Reisgen, U.; Olschok, S.; Jakobs, S.; Turner, C. Laser beam welding under vacuum of high grade materials. *Weld. World* **2016**, *60*, 403–413, doi:10.1007/s40194-016-0302-3.

13. Unt, A.; Poutiainen, I.; Grunenwald, S.; Sokolov, M.; Salminen, A. High Power Fiber Laser Welding of Single Sided T-Joint on Shipbuilding Steel with Different Processing Setups. *Appl. Sci.* **2017**, *7*, 1276, doi:10.3390/app7121276.

14. Chang, B.; Yuan, Z.; Pu, H.; Li, H.; Cheng, H.; Du, D.; Shan, J. A Comparative Study on the Laser Welding of Ti6Al4V Alloy Sheets in Flat and Horizontal Positions. *Appl. Sci.* **2017**, *7*, 376, doi:10.3390/app7040376.

15. Wang, K.; Chang, B.; Chen, J.; Fu, H.; Lin, Y.; Lei, Y. Effect of Molybdenum on the Microstructuresand Properties of Stainless Steel Coatings by Laser Cladding. *Appl. Sci.* **2017**, *7*, 1065, doi:10.3390/app7101065.

applied
sciences

MDPI

*Review*

# Highly Doped Phosphate Glass Fibers for Compact Lasers and Amplifiers: A Review

**Nadia Giovanna Boetti** [1,*], **Diego Pugliese** [2], **Edoardo Ceci-Ginistrelli** [2] , **Joris Lousteau** [3], **Davide Janner** [2] **and Daniel Milanese** [2,4]

[1]   Istituto Superiore Mario Boella, via P. C. Boggio 61, 10138 Torino, Italy
[2]   DISAT—Politecnico di Torino and INSTM, C.so Duca degli Abruzzi 24, 10129 Torino, Italy;
      diego.pugliese@polito.it (D.P.); edoardo.ceciginistrelli@polito.it (E.C.-G.); davide.janner@polito.it (D.J.);
      daniel.milanese@polito.it (D.M.)
[3]   Optoelectronics Research Centre, University of Southampton, Southampton SO17 1BJ, UK;
      J.Lousteau@soton.ac.uk
[4]   Consiglio Nazionale delle Ricerche, Istituto di Fotonica e Nanotecnologie, Via alla Cascata 56/C,
      38123 Trento, Italy
*     Correspondence: boetti@ismb.it; Tel.: +39-011-227-6312

Received: 27 October 2017; Accepted: 11 December 2017; Published: 13 December 2017

**Abstract:** In recent years, the exploitation of compact laser sources and amplifiers in fiber form has found extensive applications in industrial and scientific fields. The fiber format offers compactness, high beam quality through single-mode regime and excellent heat dissipation, thus leading to high laser reliability and long-term stability. The realization of devices based on this technology requires an active medium with high optical gain over a short length to increase efficiency while mitigating nonlinear optical effects. Multicomponent phosphate glasses meet these requirements thanks to the high solubility of rare-earth ions in their glass matrix, alongside with high emission cross-sections, chemical stability and high optical damage threshold. In this paper, we review recent advances in the field thanks to the combination of highly-doped phosphate glasses and innovative fiber drawing techniques. We also present the main performance achievements and outlook both in continuous wave (CW) and pulsed mode regimes.

**Keywords:** fiber laser and amplifier; phosphate fiber; all-fiber master oscillator power amplifier (MOPA); rare-earth-doped fiber

## 1. Introduction

Fiber lasers are nearly as old as the glass laser itself. The first demonstration of a fiber laser dated back to the 1960s, when E. Snitzer and C. J. Koester demonstrated laser action in a neodymium-doped fiber [1,2].

Nowadays, fiber lasers are not only one of the fastest growing photonic technologies, but they are also a sound industrial reality that is rapidly eroding the markets shares of other laser technologies. They find applications in a variety of fields ranging from industrial material processing, medical applications, semiconductor device manufacturing, remote sensing and scientific instrumentation.

The key advantages of fiber laser reside in their waveguide geometry. Indeed, a low threshold and a high conversion efficiency are possible thanks to the tight confinement of the pump and the generated signal. First of all the tight confinement of both pump and laser light results in a low threshold and a high optical conversion efficiency. Moreover, optical fibers exhibit an enhanced heat dissipation capability thanks to their high surface-to-volume ratio and the distribution of the thermal load over a considerably long length, which facilitates unprecedented power scaling capacity. The whole laser resonator can be realized using fiber components only, such as Fiber Bragg Gratings (FBGs) and fiber couplers. This all-fiber configuration avoids the use of free-space optics, thus allowing a robust and compact system design that facilitates the usability of fiber lasers outside the laboratory. Another advantage of fiber laser is that beam handling and beam deliveries become inherently simple thanks to the fiber based configuration [3,4].

Another important aspect of fiber laser technology is the fact that is extremely versatile and can be adapted to different laser applications, by customizing the output power, beam quality and dimension, spectral properties, output stability and temporal output properties. Moreover, the very large spectral bandwidths achievable from rare-earth (RE) ions in glass allow fiber lasers to operate from the continuous wave (CW) regime down to pulse durations of few femtoseconds.

The fiber geometry is indeed the origin of the main advantages of fiber lasers; however, it is also responsible for the main drawbacks of such devices. The enforcement of deleterious nonlinear effects, such as self-phase modulation (SPM), stimulated Raman scattering (SRS) and stimulated Brillouin scattering (SBS), is due to the confinement of the laser radiation in a small area and to the long interaction length. These effects constitute the main restriction of RE-doped fiber laser systems [3].

Optical fibers for lasers and amplifiers are drawn from glasses and several glass systems suitable for photonic applications are well-known, with a variety of chemical, physical and optical properties [5]. However, most of the research work performed so far on fiber lasers has been performed on doped silica glasses, thanks to their outstanding properties. Silica is featured by the lowest propagation losses (0.2 dB/km), can withstand high temperatures and shows exceptional mechanical strength and chemical durability. However, a major drawback of silica glass is the poor solubility of RE ions, and the consequent tendency to cluster. This causes deleterious effects on the spectroscopic properties of the material.

Multicomponent phosphate glasses have demonstrated in last years to be a promising contender to silica glass as a fiber material, especially for the realization of compact active devices [6–8]. The main reason is that, thanks to the very open and disordered matrix structure [9,10], they can withstand very high doping level of rare earth-ions (up to $10^{21}$ ions/cm$^3$), typically about 50 times higher than in the more rigid glass matrix of silica glass. High concentrations of laser-active RE ions (like Erbium, Ytterbium and Neodymium) can thus be incorporated into the glass matrix without clustering, which is responsible to degrade the performance of the glass enhancing quenching effects [11–14]. Fiber laser length can be then substantially reduced, minimizing nonlinearities that grow with fiber length.

Moreover, phosphate glasses are substantially immune to photodarkening. It was demonstrated that the maximum concentration of Yb$^{3+}$ ions allowed in a phosphate fiber before the onset of photodarkening at 660 nm was at least 56 times higher than a standard commercial silica fiber and 6 times higher than a highly Al-doped silica fiber [15].

In terms of thermo-mechanical properties, phosphate glasses have typically a low glass transition temperatures (400–700 °C), if compared to silica (1000–1200 °C), and also lower softening temperatures (500–800 °C instead of 1500–1600 °C) [10,16–18]. Although these thermal features present some advantages in terms of glass processing for fabricating the fiber preform, the drawback is that phosphate fiber can suffer from thermal degradation under high-power operation. Special care and subsequent cooling apparatus may therefore be required during high power laser operation. In addition, the characteristic temperatures difference between phosphates and silica, as well as the sharp difference in physical properties between standard commercial silica based fibers and phosphate ones, has slowed down the exploitation of these fibers due to the more challenging integration in all-fiber devices. Nonetheless, splicing of phosphate fibers to silica based fibers and fiber components (for example couplers and Fiber Bragg Gratings) is nowadays commonly employed, with relatively low loss (0.2 dB per joint) and high strength [19–21].

In the last decades, remarkable results in the field of phosphate based fiber lasers and amplifiers have been obtained in terms of high output power, new operating wavelengths, single-frequency operation, pulse duration and compact configurations. The aim of this paper is to review major results obtained in the field of phosphate fiber laser technology to highlight its complementarity to silica fiber technology and discuss its future prospect. The paper is organized as follows. In Section 2 a brief review of spectroscopic properties of main RE ions used for the development of phosphate fiber lasers and amplifiers is presented. Section 3 reports the current state of the art of CW phosphate fiber lasers and amplifiers. In Section 4 main results of CW single-frequency fiber lasers and amplifiers are reviewed. Pulsed single-frequency phosphate fiber lasers and amplifiers are addressed in Section 5.

## 2. Rare-Earth Active Ions for Phosphate Fiber Lasers

Among the different RE ions, Ytterbium is one of the most commonly used for the fabrication of phosphate fiber lasers. The electronic transition $^2F_{5/2} \rightarrow {}^2F_{7/2}$ (see Figure 1) can generate laser emission over a band of wavelengths ranging from ~970 to ~1200 nm, thus allowing a wide tuning ability of the output beam in the 1 µm window and short pulse amplification (down to fs regime) [22].

The absorption band is also very broad, from ~850 to ~1070 nm, and covers a range where laser diodes have their highest output power and efficiency. Since the pump wavelength is very close to the laser emission one, an $Yb^{3+}$-doped fiber laser displays always a small quantum defect, potentially allowing for very high power efficiencies and reduced overheating effects in high-power regimes [23].

Ytterbium is well-known for the simplicity of its energy level diagram, based on a simple two levels system that reduces the incidence of detrimental processes such as multiphonon relaxation and excited state absorption (ESA). Figure 1a shows a typical $Yb^{3+}$ ions energy level diagram with indicative splitting of the Stark sublevels. This splitting allows the ion to work in a three or four levels system, according to the choice of the pump and lasing wavelengths. Figure 1b displays typical absorption and emission spectra of a multicomponent $Yb^{3+}$-doped phosphate glass. Table 1 summarizes the typical absorption and emission cross-sections reported in literature, alongside typical lifetime value of the $^2F_{5/2}$ excited state level.

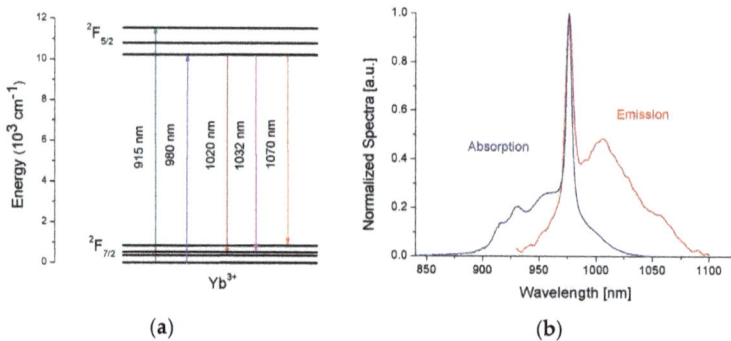

(a)  (b)

**Figure 1.** (a) Typical energy level diagram of $Yb^{3+}$ ions in phosphate glasses; (b) Typical absorption and emission spectra of $Yb^{3+}$ ions in an in-house fabricated phosphate glass.

The Neodymium is featured by a much more complex structure. Figure 2a reports the schematic of the energy levels of $Nd^{3+}$ ions with the transitions of interest with respect to this review.

Although the radiative decay can occur to all the lower lying $^4I_J$ levels, the strongest transition is the one to the $^4I_{11/2}$ level, that leads to a maximum peak at a wavelength of 1053 nm in most phosphate glasses. Thanks to the numerous absorption bands in the visible and near-infrared, Nd could be pumped by flash lamps, which were the oldest and most common pumping solution for this ion in phosphate glass host. The first fiber laser in the 1960's was indeed demonstrated in a Nd-doped fiber excited by means of flash lamps [1,2]. More recently, with the availability of laser diodes at the requested wavelengths, a more efficient and compact pumping scheme has been achieved through the use of AlGaAs pump diode lasers operating at around 808 nm to excite the transition $^4I_{9/2} \rightarrow {}^2H_{9/2} + {}^4F_{5/2}$ (see Figure 2a). Despite the large number of closely spaced excited states, $Nd^{3+}$ ion does not suffer from ESA. Even though the $^4F_{3/2} \rightarrow {}^2D_{5/2}$ ESA transition occurs, it is resonant with the photon energy of the 800 nm wavelength laser pump, it is negligible and does not significantly affect the 800 nm pumping [24]. Thanks to the several absorption bands in the visible, an unusual pumping scheme for Nd-doped phosphate lasers by solar radiation was also investigated in the framework of the research on solar pumped lasers [25,26].

Figure 2b reports typical absorption and emission spectra of $Nd^{3+}$ ions in phosphate glasses. Typical cross-section values are reported in Table 1, along with lifetime of the $^4F_{3/2}$ excited state in phosphate glass.

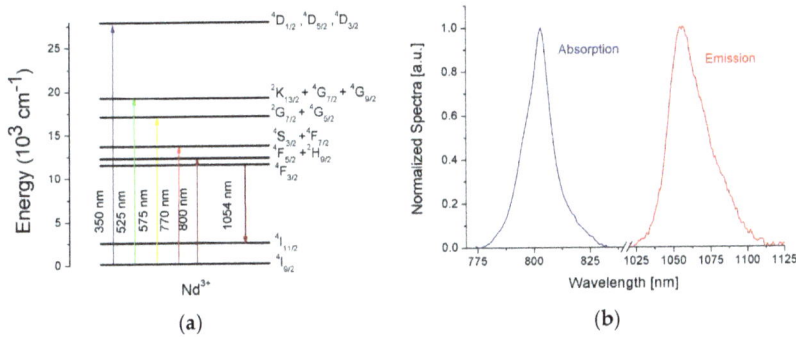

**Figure 2.** (a) Typical energy level diagram of $Nd^{3+}$ ions in phosphate glasses; (b) Typical absorption and emission spectra of $Nd^{3+}$ ions in phosphate glasses (adapted from [27]).

The most common laser transition in Erbium is that from the $^4I_{13/2}$ manifold to the ground-state $^4I_{15/2}$ (see Figure 3a), centered at around 1.55 μm. This transition forms a three-levels system for which a high pump power is required to obtain a population inversion and thus reach the threshold. The most common pumping arrangement is based on the transition $^4I_{15/2} \rightarrow {}^4I_{11/2}$ with a wavelength of around 980 nm, although in-band pumping ($^4I_{15/2} \rightarrow {}^4I_{13/2}$, e.g., at 1.45 μm) is also possible [28,29]. This transition is widely exploited in $Er^{3+}$-doped fiber amplifiers (EDFAs).

In order to increase lasers and amplifiers efficiency, Erbium-doped fibers and amplifiers are often co-doped with Ytterbium, which exhibits larger absorption cross-section and broad absorption band (from ~850 to ~1070 nm). In this scheme Ytterbium, which acts as a sensitizer, strongly absorbs pump photons, and then resonantly transfers its energy to Erbium that acts as activator.

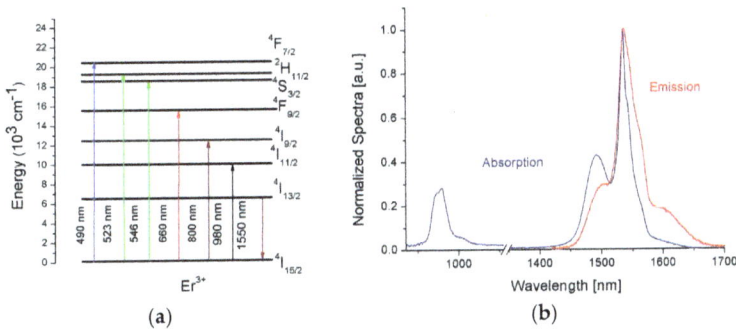

**Figure 3.** (a) Typical energy level diagram of $Er^{3+}$ ions in phosphate glasses; (b) Common absorption and emission spectra of $Er^{3+}$ ions in phosphate glasses (adapted from [12]).

Phosphate glass is considered an excellent host for an $Yb^{3+}$-$Er^{3+}$ co-doped system in virtue of its high emission cross-section, low accumulative energy transfer rates and low backward energy transfer [30]. In fact, in comparison to silicate glasses ($\sim$1100 cm$^{-1}$), the larger phonon energy in the phosphate host ($\sim$1200 cm$^{-1}$) increases the transition probability for $^4I_{11/2} \to {}^4I_{13/2}$ relaxation [31], which prevents the back energy transfer from $Er^{3+}$ to $Yb^{3+}$. Furthermore, thanks to the large spectral overlap between the $Yb^{3+}$ emission spectrum ($^2F_{5/2} \to {}^2F_{7/2}$) and $Er^{3+}$ absorption spectrum ($^4I_{15/2} \to {}^4I_{11/2}$), the energy transfer efficiency from $Yb^{3+}$ to $Er^{3+}$ in phosphate glasses can reach 95% [11].

Typical absorption and emission cross-sections of Erbium ions in phosphate glasses, together with lifetime of the $^4I_{13/2}$ level, are reported in Table 1.

**Table 1.** Typical spectroscopic parameters of common rare-earth (RE) ions in highly doped phosphate glasses [8,11,14,25,32–35].

| Parameter | Ytterbium | Neodymium | Erbium |
|---|---|---|---|
| Absorption cross-section (cm$^2$) | $\sim$1.6 $\times$ 10$^{-20}$ $^2F_{7/2} \to {}^2F_{5/2}$ | $\sim$2.0 $\times$ 10$^{-20}$ $^4I_{9/2} \to {}^2H_{9/2} + {}^4F_{5/2}$ | $\sim$1.5 $\times$ 10$^{-21}$ $^4I_{15/2} \to {}^4I_{11/2}$ $\sim$6.6 $\times$ 10$^{-21}$ $^4I_{15/2} \to {}^4I_{13/2}$ |
| Emission cross-section (cm$^2$) | 1.4 $\times$ 10$^{-20}$ $^2F_{5/2} \to {}^2F_{7/2}$ | $\sim$2.4 $\times$ 10$^{-20}$ $^4F_{3/2} \to {}^4I_{11/2}$ | $\sim$7.0 $\times$ 10$^{-21}$ $^4I_{13/2} \to {}^4I_{15/2}$ |
| Doping level (ions/cm$^3$) | 1.4 $\times$ 10$^{21}$ | $\sim$2.7 $\times$ 10$^{20}$ | 4.0 $\times$ 10$^{20}$ |
| Excited state lifetime (ms) | 1.2 $^2F_{5/2} \to {}^2F_{7/2}$ | 0.3 $^4F_{3/2} \to {}^4I_{11/2}$ | 8.0 $^4I_{13/2} \to {}^4I_{15/2}$ |

## 3. CW Phosphate Fiber Lasers and Amplifiers

RE-doped glass fibers are excellent gain media for obtaining high power lasers with enhanced brightness. The large surface-to-volume ratio of active fibers favors heat dissipation to the surrounding, allowing active fibers to operate at very high powers. The heat dissipation further increases by using efficient RE dopants, such as Ytterbium, which allows operation with pump wavelengths close to the signal wavelength [23]. As a consequence, beam distortion is negligible and the beam quality is mainly affected by the fiber geometry.

In addition, the waveguide structure confines the pump light over the entire fiber length, enabling noticeably large single-pass gains to be achieved due to the long interaction path of the light with the active medium. This leads to a very efficient operation of the fiber lasers, showing very high gain and low pump threshold values.

Nowadays, high power fiber lasers and amplifiers are typically implemented using RE-doped double-cladding (DC) optical fibers, a concept which was introduced in 1988 by Snitzer et al. [36] and has allowed in last decades a significant enhancement in scaling fiber lasers to higher power levels. DC fibers consist in a RE-doped core surrounded by a much larger inner cladding, itself surrounded by an outer cladding (see Figure 4). The pump light is launched and confined into the inner cladding and spatially overlaps the core. The pump power is then absorbed over the entire length of the fiber through multiple interaction with the core, thus exploiting the fiber geometry. The benefit of this design resides in the possibility to use multi-mode high power laser diodes with relatively poor beam quality and to obtain a single-mode fiber radiation with outstanding beam quality.

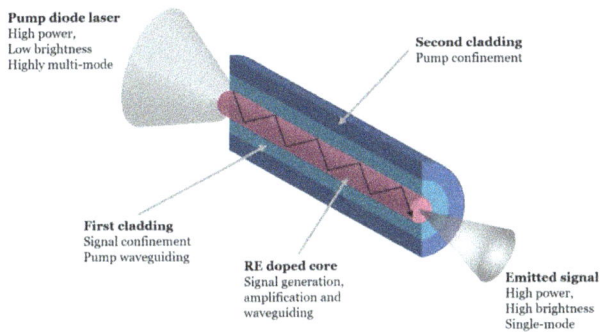

**Figure 4.** Schematic of cladding-pumped double-cladding fiber laser.

Thanks to DC fiber geometry, power scaling of silica-based fiber lasers has experienced noticeable progress in the last two decades, reaching 100 kW of output power in the 1 μm wavelength region from a multi-mode $Yb^{3+}$-doped fiber [37] and 10 kW from a single-mode fiber laser [38]. In eye-safe region, at 1.5 μm, 297 W from an $Yb^{3+}$-$Er^{3+}$-doped fiber laser was demonstrated [39], while at 2 μm 1 kW of emitted power was reported from a $Tm^{3+}$-doped fiber laser [40].

Power scaling of silica fiber lasers is ultimately limited by two main mechanisms: nonlinear effects (especially SBS) induced by the long path lengths of the tightly confined light in the fiber core and photodarkening. A different approach devoted to the fiber laser power scaling, which can overcome the above limitations, is indeed to employ phosphate glass instead of silica as fiber material. The higher solubility of RE ions in phosphate glass hosts allows the increase of RE ions concentration and thus to generate the necessary amplification over a much shorter fiber length as compared to silica fibers. The onset of the non-linear phenomena described above is then considerably limited. In addition, the higher photodarkening threshold in phosphate glasses eliminates the risk of power degradation over time in high power applications [15].

Another important advantage of phosphate glasses in power scaling of fiber lasers is related to the possibility to easily adjust the refractive indices [41]. Thus, DC optical fibers with a reduced core numerical aperture (NA) and hence a much larger single-mode core are feasible. Furthermore, thanks to a large glass forming region, it is possible to manufacture glasses with considerably large refractive index differences and thus realize a high NA pump waveguide with a glass based outer cladding instead of a low-index polymer coating, as widely used in DC silica fibers. These all-glass DC optical fibers offer a high degree of immunity to heat loading [8,42].

A typical configuration used to realize high power fiber lasers is the master oscillator power amplifier (MOPA) system: the high power laser is obtained by splitting it into a low-power signal generator (master oscillator) and an optical power amplifier that boosts the output power. Thus, the high-spectral qualities of the signal are preserved while simplifying the manufacturing constraints of the seed-laser. For this kind of device, an all-fiber structure without the use of free-space components is a better solution in order to simplify configurations while improving compactness and reliability.

Phosphate glasses main features, such as extremely high doping level, low SBS gain coefficient and absence of photodarkening, make them an ideal host to develop a MOPA system.

*3.1. Yb-Doped Phosphate Fiber Lasers and Amplifiers*

The first CW cladding-pumped Yb-doped single-mode phosphate fiber laser emitting in the 1 μm region was demonstrated in 2006 by Lee et al. [8] using a Large Mode Area (LMA) fiber with diameters of 10/240/355 μm for the core, inner cladding and outer cladding, respectively. In view of breaking the symmetry of the fiber design and thus increasing the pump absorption, the core of the fiber was offset from the center and an air hole parallel to the core was added to the inner cladding. With a length of 84.6 cm of this fiber, doped with 12 wt % of $Yb_2O_3$ and pumped with 80 W at 940 nm, an output power of nearly 20 W at 1.07 μm, with a slope efficiency of 25.8% against launched pump power, was demonstrated. With this very same fiber configuration, few years later, power scaling to 57 W was demonstrated with a slope efficiency of 50.6% by varying the pump wavelength to 977 nm, where $Yb^{3+}$ absorption is much higher and quantum defect is lower [43]. In both reported Yb-doped fiber lasers, the slope efficiency was reduced by the quite high loss of the fiber, around 3 dB/m, which was measured to be mainly due to impurity absorption (77%) and scattering.

Recently, an Ytterbium-doped phosphate glass single-mode fiber laser with noticeably high slope efficiency of 66.6% in relation to the launched pump power was reported. The fiber consisted in a double-cladding fiber with a 19 μm diameter step-index core and a pump waveguide with a high numerical aperture due to the external air-cladding. An output power of 11.6 W was obtained by using 4 cm-long fiber sample pumped with a multi-mode laser diode. This is the best result, in terms of slope efficiency, ever reported for a phosphate single-mode fiber laser pumped with a multi-mode laser diode [44].

Concerning $Yb^{3+}$-doped phosphate fiber amplifiers, the first watt level $Yb^{3+}$-doped phosphate fiber MOPA was reported in 2009 [43] with an output power of 16.3 W corresponding to a gain of 27 dB with respect to the launched signal power (~0.36 dB/cm). The gain medium was a 74.5 cm-long single-mode double-cladding phosphate fiber doped with 12 wt % of $Yb_2O_3$.

### 3.2. Nd-Doped Phosphate Fiber Lasers and Amplifiers

As the first fiber laser was realized using a $Nd^{3+}$-doped fiber, in that case made of silicate glass [1], the first phosphate fiber laser reported in literature was also manufactured with a $Nd^{3+}$-doped phosphate fiber [45].

A length of 10 mm of single-mode and single-cladding optical fiber, manufactured by rod-in-tube technique and highly doped with 3 wt % of $Nd_2O_3$ ($3.1 \times 10^{20}$ $Nd^{3+}$ ions/$cm^3$), was used to demonstrate laser emission at 1.054 μm, when core-pumped by an 808 nm laser diode. A low threshold of 1 mW and an efficiency of ~50% were measured. Lasing was also demonstrated at 1.366 μm, although with lower performance both in threshold and slope efficiencies, as expected for this less efficient neodymium emission.

In 1996 Griebner et al. [46] reported a core-pumped Nd-doped multi-mode phosphate fiber, although with a quite low output power of 130 mW at 1053 nm. An efficiency of 31% with respect to the launched pump power was achieved at 807 nm.

Concerning power scaling of CW $Nd^{3+}$-doped phosphate fiber lasers, in 2011 Zhang et al. [47] reported an output power of 2.87 W at 1053 nm from a length of 26 cm of heavily $Nd^{3+}$-doped ($3.5 \times 10^{20}$ $Nd^{3+}$/$cm^3$) DC fiber, with diameters of 14/285/360 μm for the core, inner cladding and outer cladding, respectively. Average slope efficiencies of 26.2% and 44.7% with respect to the launched and absorbed pump powers at 795 nm were obtained. The reported loss of 1.5 dB/m at 1053 nm is the lowest ever measured for a $Nd^{3+}$-doped phosphate fiber.

More recently [48], single-mode operation was reported in a cladding-pumped $Nd^{3+}$-doped fiber laser having an output power of 1.42 W and an efficiency of 34% with respect to the absorbed pump power at 808 nm. A 21 cm-long DC fiber was used, with a small core diameter of 5.3 μm to allow single-mode operation at 1.053 μm. In view of improving the coupling efficiency between the pump diode and the active fiber, a novel design for local cooling by fluid sealing of the coupling point was exploited.

In 2015 a $Nd^{3+}$-doped phosphate fiber laser was demonstrated by using a DC fiber with an hexagonal inner cladding made by a stack-and-draw technique, that is commonly used for the realization of microstructured optical fibers [42]. Single-mode operation was achieved with a core diameter of 35 μm, thanks to the low NA of the fiber core. A maximum output power of 2.87 W and a slope efficiency of 19% were achieved. The poor laser performance was due to the high propagation loss of the fiber, as high as 3.8 dB/m, and to the mismatch between the pump wavelength (793 nm) and the Nd-doped glass peak absorption (802 nm).

An alternative solution that was recently proposed for the power scaling of $Nd^{3+}$-doped fiber lasers and amplifiers relies in the use of short-length core/cladding canes as active elements. A 60 mm length of cane, with a core diameter of 270 μm and a cladding diameter of 800 μm, was able to emit 2 W of output power at 1054 nm with slope efficiencies of 30% and 40% with respect to the launched and absorbed pump powers, respectively. Caird's analysis revealed that a slope efficiency of 55% over the absorbed power could be achieved in optimal conditions, matching the highest results obtained so far [27]. Although the latter configuration did not operate in single-mode regime, similar cane configuration can be employed as a booster amplifier in a MOPA configuration to reach high power while preserving high beam quality [49].

Significant results have been obtained in last years from photonic crystal fibers (PCFs). In 2012 the first non-silica all-solid $Nd^{3+}$-doped PCF was demonstrated, with an hexagonal shaped inner cladding, to achieve high absorption coefficient of pump power [50]. An output power of 7.92 W and a slope efficiency of 38.1% were obtained but with unsatisfactory beam quality. In 2014 a single-mode multi-watt phosphate $Nd^{3+}$-doped all-solid PCF with core diameters of 30, 35 and 40 μm was reported [51]. An output power of 5.4 W and a slope efficiency of 31% were measured with a beam quality that is not degraded even at the highest power.

Higher emitted power was registered by using a complex photonic crystal fiber (PCF) containing seven different single-mode cores. This allowed achieving 15.5 W of emitted power over a fiber length of 25 cm, featuring a slope efficiency of 57% over the absorbed power [52].

*3.3. Er and Yb/Er Co-Doped Phosphate Fiber Lasers and Amplifiers*

Erbium-Ytterbium co-doped system received noticeable interest in the last two decades due to its use for optical communications and eye-safe laser applications.

In 2004 Qiu et al. reported 9.3 W CW 1535 nm multi-mode output from a 7.0 cm short-length single-cladding Er-Yb phosphate fiber [7]. The fiber featured a D-shaped cladding and an off-centered core 130 and 20 μm in diameter, respectively. A slope efficiency of 29% and a power per unit length (p.u.l.) as high as 1.33 W/cm were achieved. From another 7.1 cm of Er-Yb co-doped fiber laser, 4.0 W single transverse mode output was generated with lower power per unit fiber length (0.56 W/cm), due to the less efficient pump absorption exhibited by the smaller core.

In the early 2000's the first $Er^{3+}$-doped fiber amplifiers based on heavily Erbium-doped phosphate fibers were demonstrated [53,54]. They were core-pumped by single-mode laser diodes emitting at 980 nm. For example, Hwang et al. [54] reported a high concentration (3.5 wt %) $Er^{3+}$-doped phosphate fiber amplifier. An internal gain of 23 dB and a gain p.u.l. of 3 dB/cm were achieved in a 71 mm-long $Er^{3+}$-doped phosphate fiber core-pumped by a laser diode at around 980 nm. The fiber, manufactured by the rod-in-tube technique, displayed a single-mode behavior at 1.55 μm and showed a quite high loss of 0.28 dB/cm at 1.3 μm. The gain obtained was lower than that predicted by theoretical calculations, due to cooperative up-conversion phenomena between the $Er^{3+}$ ions and the small absorption cross-section of Erbium at the pumping band of 980 nm. Thus, soon after the same group proposed the realization of the amplifier using a phosphate fiber doped with a lower amount of $Er^{3+}$ (3 wt %) [55], to prevent Erbium concentration quenching, and with $Yb^{3+}$ sensitization (2 wt %), to increase the pump absorption. An internal gain of 18 dB and a gain p.u.l. of 5 dB/cm were obtained for small-signal input at 1535 nm, using 3.6 cm-long co-doped fiber pumped at 980 nm in a co-propagating configuration.

In light of limiting the amplifier cost and power scale the output power, single-cladding phosphate fibers were replaced by DC fibers pumped by broad area high power multi-mode laser diodes instead of single-mode pump sources. Thanks to these improvements, in 2003 an $Er^{3+}$-doped phosphate fiber amplifier showing net gain values of 41, 27 and 21 dB at 1535 nm, 1550 nm and over the C-band, respectively, was demonstrated with only 8 cm-long fiber. The fiber was an all-glass single-mode DC fiber with a loss of 0.09 dB/cm at 1.3 μm [6], which was significantly lower compared to the previously fabricated fibers.

In 2008 5 cm-long single-mode heavily $Yb^{3+}/Er^{3+}$ co-doped fiber was used to demonstrate optical amplification of signal in the range 1525–1565 nm [56]. The fiber, doped with 1 mol% of $Er^{3+}$ and 2 mol% of $Yb^{3+}$, was dual-pumped by two 976 nm fiber pigtailed laser diodes. A peak net gain of 16.5 dB at 1534 nm was achieved, that corresponds to a gain p.u.l. of 3.3 dB/cm (internal gain p.u.l. of 9.1 dB/cm). The fiber loss was evaluated to be 0.06 dB/cm at 1.3 μm.

In 2015 novel high concentration Yb/Er co-doped phosphate glasses (1.08 mol% of $Er_2O_3$ and 1.08 mol% of $Yb_2O_3$) were employed to manufacture DC fibers for optical amplification. Core-pumped and cladding-pumped amplifiers were demonstrated, reaching gains p.u.l. of 4.0 and 2.3 dB/cm, respectively [57].

The main results obtained with highly RE-doped CW phosphate fiber lasers are summarized in Table 2.

**Table 2.** Main results obtained with continuous wave (CW) fiber lasers based on highly RE-doped phosphate glasses.

| Fiber Type [Reference] | RE Doping Concentration | $\lambda_{pump}$ (nm) | $\lambda_{signal}$ (nm) | Output Power (W) | Slope Efficiency (%) | Gain Length (cm) |
|---|---|---|---|---|---|---|
| $Yb^{3+}$-doped multi-mode [8] | 12 wt % | 940 | 1064 | 20 | 34.4 | 84.6 |
| $Yb^{3+}$-doped single-mode [43] | 12 wt % | 977 | 1064 | 57 | 56.7 | 71.6 |
| $Nd^{3+}$-doped multi-mode [47] | $3.5 \times 10^{20}$ $Nd^{3+}/cm^3$ | 795 | 1053 | 2.9 | 44.7 | 26 |
| $Nd^{3+}$-doped single-mode [48] | $3.5 \times 10^{20}$ $Nd^{3+}/cm^3$ | 808 | 1053 | 1.4 | 34.1 | 21 |
| $Nd^{3+}$-doped single-mode PCF [52] | 3 wt % | 793 | 1053 | 15.5 | 57 | 25 |
| $Yb^{3+}/Er^{3+}$ co-doped multi-mode [7] | $8.6 \times 10^{20}$ $Yb^{3+}/cm^3$ $1.1 \times 10^{20}$ $Er^{3+}/cm^3$ | 976 | 1535 | 9.3 | 39 | 7.0 |
| $Yb^{3+}/Er^{3+}$ co-doped single-mode [7] | $8.6 \times 10^{20}$ $Yb^{3+}/cm^3$ $1.1 \times 10^{20}$ $Er^{3+}/cm^3$ | 976 | 1535 | 4.0 | 39 | 7.1 |

## 4. CW Single-Frequency Phosphate Fiber Lasers and Amplifiers

Fiber lasers considered in the previous Sections display relatively broad linewidth, usually in the 1–10 nm range. However, several applications such as interferometric gravitational-wave detection [58], coherent Light Detection and Ranging (LIDAR) [59], laser nonlinear frequency conversion [60], coherent beam combining [61] and laser spectroscopy [62] require laser emissions with narrower linewidth. This feature can be provided by single-frequency fiber lasers, which are lasers that operate with only a single-longitudinal mode and thus can emit quasi-monochromatic radiation with a very narrow linewidth and low noise.

Different configurations have been exploited to demonstrate single-longitudinal mode operation. One typical design is the distributed feedback (DFB) fiber laser, in which a FBG is directly written in the core of the active fiber and is responsible for the generation of a phase change in the middle of the grating area.

Another common configuration is the distributed Bragg reflector (DBR) fiber laser, in which the laser resonator is formed by the RE-doped fiber placed between two FBGs. This configuration is generally preferred for the fabrication of single-frequency fiber lasers thanks to its compactness and robustness [63].

When an output power higher than 1 W is demanded, a MOPA configuration is used, with a single-frequency seed laser followed by a series of amplifier stages to achieve the power amplification. This configuration is able to power scale the output power while preserving the spectral characteristics of the seed laser.

The use of highly doped phosphate glass fibers for the manufacturing of single-frequency fiber lasers is particularly advantageous since allows the realization of a short cavity that creates large longitudinal mode spacing, helping to maintain lasing on a single-longitudinal mode and thus a single-frequency behavior.

Power scaling of single-frequency fiber lasers is ultimately reduced by SBS and is much more arduous than that of a multi-longitudinal mode CW fiber laser. Several different methods have been implemented to limit the SBS in silica-based fiber lasers [64–66]. The use of phosphate glass-based optical fiber represents an alternative option to overcome this limitation.

CW single-frequency phosphate-based fiber lasers operating at 1 and 1.5 μm have reached a relatively high maturity level and are commercially available [67].

## 4.1. Yb-Doped Single-Frequency Phosphate Fiber Lasers

In 2004 Kaneda et al. [68] reported for the first time a DBR laser based on 1.5 cm-long $Yb^{3+}$-doped phosphate fiber that was fusion spliced with two FBGs to form the resonator. 200 mW of single-frequency output power at 1064.2 nm was achieved (gain p.u.l. equal to 2 dB/cm), with a linewidth of approximately 3 kHz.

In 2011 Xu et al. [69] demonstrated over 400 mW of emission at 1063.90 nm from a DBR $Yb^{3+}$-doped phosphate fiber laser based on 0.8 cm-long active fiber. The fiber was heavily doped (15.2 wt %) in order to obtain a very short resonator cavity and thus ensure laser operation on single-longitudinal mode. A slope efficiency of about 73% versus launched pump power was obtained, with a net gain coefficient p.u.l. of 5.7 dB/cm. This work reports the highest slope efficiency and output power obtained so far in this kind of single-frequency fiber lasers.

Besides the typical lasing wavelength at 1064 nm, $Yb^{3+}$-doped phosphate fibers have been successfully exploited to obtain single-frequency operation at more exotic wavelengths [20,70,71], thanks to the broad $Yb^{3+}$ emission from 970 to 1200 nm.

In particular, during last years, interest for high power single-frequency lasers below 1 μm has increased, since they are excellent pump sources for second and fourth harmonic generation in the blue and deep UV, which have found several scientific and industrial applications [72]. $Yb^{3+}$-doped phosphate fibers have been successfully employed to achieve single-frequency laser emitting at 976 nm, through the transition from the lowest level of the excited state $^2F_{5/2}$ manifold to the lowest level of the ground state $^2F_{7/2}$ manifold (quasi-three-level operation) (see Figure 1a). $Yb^{3+}$-doped phosphate fibers, compared to $Yb^{3+}$-doped silica fibers, exhibit a larger difference between the absorption and emission cross-sections at shorter wavelengths, making them more suitable for short wavelength laser emission [73].

The first demonstration of an $Yb^{3+}$-doped phosphate fiber laser emitting at 976 nm was reported by Bufetov et al. [32] in 2006 using 2 cm-long highly doped fiber ($1 \times 10^{21}$ $Yb^{3+}$ ions/cm$^3$). 250 mW of output power was achieved, with an efficiency versus launched pump power of about 40%.

In 2012 the first 976 nm single-frequency all-fiber DBR fiber laser was developed with a 2 cm-long highly (6 wt %) $Yb^{3+}$-doped phosphate fiber [20]. 100 mW of linearly polarized output power was generated with a linewidth of less than 3 kHz. By using this fiber laser as a seeder in a MOPA configuration, a single-frequency $Yb^{3+}$-doped phosphate fiber amplifier was realized. A linearly polarized output power of 350 mW was obtained by core-pumping a 4 cm-long polarization maintaining (PM) phosphate fiber [74]. The birefringence of the PM fiber was made possible by inserting two high stress elements around the fiber core. A small signal net gain of 25 dB, corresponding to a unit gain over 6 dB/cm, was achieved.

With the aim to investigate the power scaling of a single-frequency laser operating at 976 nm, Wu et al. fabricated an $Yb^{3+}$-doped phosphate fiber amplifier based on 7 cm-long $Yb^{3+}$-doped DC phosphate fiber cladding-pumped by high power multi-mode laser diodes. 3.41 W single-frequency output power was obtained, with a relatively low slope efficiency of ~7% due to the reduced spatial overlap of the pump and the doped fiber core in the cladding-pumped arrangement [73].

Single-frequency emission from an $Yb^{3+}$-doped phosphate fiber was also demonstrated at 1014 nm, which is an interesting wavelength for optical lattice clock applications [75]. A compact fiber laser of 5 mm was reported, with a CW single transverse and longitudinal mode characterized by an output power over 164 mW, low noise and a linewidth lower than 7 kHz. At the wavelength of 1014 nm Yang et al. [76] reported also a MOPA laser with over 1 W of single-frequency emission, obtained by amplifying a seed laser with a core-pumped one stage $Yb^{3+}$-doped phosphate fiber amplifier of 4 cm of length.

On the other hand, with $Yb^{3+}$-doped phosphate fibers, emission at longer wavelengths was also investigated. Low noise single-frequency single polarization fiber laser emitting at 1083 nm was reported from a 1.8 cm-long heavily (18.3 wt %) $Yb^{3+}$-doped phosphate fiber. 100 mW of output power was achieved, with a slope efficiency of almost 30% and a laser linewidth lower than 2 kHz [70].

*4.2. Er and Yb/Er Co-Doped Single-Frequency Phosphate Fiber Lasers*

The first single-frequency phosphate fiber laser emitting at 1.5 μm was reported in 2003 [77], with an output power of 100 mW and a very narrow linewidth of less than 2 kHz. It was a DBR fiber laser based on an heavily $Yb^{3+}/Er^{3+}$ co-doped fiber (3 wt % of $Yb^{3+}$ and 2 wt % of $Er^{3+}$) that displays very high optical gain per unit length of up to 5 dB/cm. The laser cavity was established by using two spectrally narrow passive FBGs that were fusion spliced to a 2 cm-long piece of active fiber.

In 2013 [78] a monolithic all-phosphate glass fiber laser was reported, with a laser cavity obtained by inscribing a FBG directly into an heavily $Yb^{3+}/Er^{3+}$ co-doped phosphate fiber (8 wt % of $Yb_2O_3$ and 1 wt % of $Er_2O_3$) by using a femtosecond laser. Avoiding hybrid splices between phosphate and silica based fibers allowed for improving the laser stability and reducing the cavity loss. An output power of 550 mW was demonstrated with a slope efficiency of 12%.

Power scaling of a single-frequency fiber laser emitting at 1.5 μm was investigated in 2005 [79] by using a DC heavily $Yb^{3+}/Er^{3+}$ co-doped phosphate fiber, cladding-pumped by multi-mode laser diodes at 976 nm. 1.6 W of single-longitudinal mode output power was reported from a 5.5 cm length of fiber, although the slope efficiency was limited at 5%.

The main results obtained with highly RE-doped CW single-frequency phosphate fiber lasers are summarized in Table 3.

**Table 3.** Main results obtained with CW single-frequency phosphate fiber lasers based on highly RE-doped phosphate glasses.

| Fiber Type [Reference] | RE Doping Concentration | $\lambda_{pump}$ (nm) | $\lambda_{signal}$ (nm) | Output Power (mW) | Slope Efficiency (%) | Gain Length (cm) | Linewidth (kHz) |
|---|---|---|---|---|---|---|---|
| $Yb^{3+}$-doped [69] | 15.2 wt % | 976 | 1064 | 400 | 72.7 | 0.8 | <7 |
| $Yb^{3+}$-doped [20] | 6 wt % | 915 | 976 | 100 | 25 | 2 | <3 |
| $Yb^{3+}$-doped [75] | 15.2 wt % | 976 | 1014 | 164 | 21.9 | 0.5 | <7 |
| $Yb^{3+}$-doped [70] | 18.3 wt % | 976 | 1083 | 100 | 29.6 | 1.8 | <2 |
| $Yb^{3+}/Er^{3+}$ co-doped [78] | 8 wt % $Yb_2O_3$ 1 wt % $E_2O_3$ | 975 | 1538 | 550 | 12 | 7 | <60 |
| $Yb^{3+}/Er^{3+}$ co-doped [79] | $8.6 \times 10^{20}$ $Yb^{3+}/cm^3$ $1.1 \times 10^{20}$ $Er^{3+}/cm^3$ | 976 | 1550 | 1600 | 5 | - | 5.5 |

## 5. Pulsed Phosphate Fiber Lasers and Amplifiers

Pulsed fiber laser operation is generally obtained through modulation of the laser cavity using several Q-switching and mode-locking techniques. Owing to the very large spectral bandwidths achievable from RE ions in glass hosts, pulsed fiber lasers can operate in nanosecond, picosecond and even femtosecond regimes using chirped pulsed amplification techniques.

Q-switching is a technique mostly applied for the generation of micro and nanosecond pulses of high energy and peak power, while the mode-locking technique enables ultrafast laser pulses in the picosecond and femtosecond ranges.

In 1993 a Q-switched fiber laser based on a multicomponent Nd-doped phosphate glass was reported with 2.2 µJ, 2 ns duration pulses at up to 2 kHz repetition rate with only 22 mW of absorbed pump power at 812 nm from a laser diode [80]. Tunability over a 40 nm tuning range around 1060 nm was demonstrated thanks to the broad emission line of Neodymium in the phosphate glass.

More recently, in 2004 [81], a single-frequency all-fiber Q-switched laser at 1550 nm was reported with 25 W of peak power in 12 ns, 0.3 µJ pulses at a repetition rate of 80 kHz with 370 mW of pump power, without using an amplifier. The laser consisted of a 2 cm-long Yb/Er co-doped phosphate glass fiber fusion spliced between two FBGs inscribed in silica fibers. The fiber was actively Q-switched by fast stress-induced birefringence modulation, a technique that allows an all-fiber configuration, thus avoiding the use of bulk components in the cavity responsible for the growth of loss, size and complexity.

By using 2 cm-long highly (6 wt %) $Yb^{3+}$-doped phosphate fiber the first compact, single-frequency all-fiber Q-switched laser at 1 µm was reported in 2007 [82]. The fiber was actively Q-switched by fast stress-induced birefringence modulation [81], a technique previously demonstrated at 1.5 µm. A repetition rate up to 700 KHz was reported, with a maximum average power of 31 mW. A maximum average pulse energy of 0.4 µJ was obtained.

Concerning fiber lasers operating in the picosecond regime, mode-locking techniques are generally used. Passive mode-locking is usually preferred to avoid costly modulators in the laser cavity as those commonly employed in active mode-locking solutions. However, passive mode-locking of fiber lasers allows achieving low repetition rates (tens of MHz), as a results of the long cavity length. Thus, an attractive approach to push the pulse repetition rate into the GHz regime is the use of short Fabry–Perot lasers with high gain RE-doped phosphate fiber and a small saturable absorber that could be integrated with fiber-end.

In 2007 [83] a mode-locked short cavity 10 GHz phosphate fiber laser emitting at 1535 nm was reported using a 1 cm-long heavily Yb/Er co-doped phosphate fiber and a saturable absorber constituted by single-wall carbon nanotubes. Stable pulse trains with an output power as high as 30 mW and a repetition rate as high as 10 GHz were achieved.

In 2012 another solution for the integrated saturable absorber was proposed: a graphene film deposited on one tip of the active fiber [84]. The obtained laser operated in a stable mode-locked regime with fundamental repetition rate of 7 GHz but with an output power of only 1.2 mW.

In 2014 [85] a repetition rate of 12 GHz was demonstrated from a compact and stable all-fiber fundamentally mode-locked fiber laser based on an Yb/Er co-doped phosphate fiber and a semiconductor saturable absorbing mirror (SESAM). A short length of 0.8 cm of high gain (6 dB/cm at 1535 nm) PM fiber was used to form the laser cavity.

Several applications like coherent LIDAR, range finding or active remote sensing require, for high precision measurements, not only high peak power laser pulses in the nanosecond time regime but also a narrow linewidth. However, neither directly Q-switching a single-frequency fiber laser nor modulating a CW single-frequency fiber laser with an external modulator can deliver sufficient energy/power for practical applications [86]. A MOPA system is thus required to boost the power of a single-frequency low power seed laser.

Power scaling of narrow linewidth fiber laser and amplifiers using single-mode silica fibers is challenging due to the onset of nonlinearities in the fiber, primarily SBS, that build up strongly in case of pulses >1 ns [87].

Therefore, increasing SBS threshold for fiber amplifiers is crucial and several solutions have been proposed to fulfill this scope [88–90]. One possible approach is to focus on the optical fiber itself, with a reduction of the active fiber length and the increase of the core size, conditions that can be achieved more easily by using active fibers made of phosphate glass.

In 2009 Shi et al. [21] proposed the use of a single-mode PM large core highly Er/Yb co-doped phosphate fiber for engineering the final power amplifier stage of a monolithic all-fiber pulsed MOPA laser operating in the C band. Single-mode operation was achieved in such a large core (15 μm) thanks to the accurate control of the refractive indices of the core and the cladding obtainable in phosphate glasses (core NA = 0.053). The fiber length was 12 cm only, thanks to the high doping level used in the phosphate glass core (15% Yb and 3% Er). Pulse energy of 54 μJ and peak power of 332 W without SBS effects for 153 ns pulses at 1538 nm were demonstrated.

In 2010 the same research group, by using 15 cm of a new active fiber with larger single-mode core (25 μm, core NA = 0.0395) for the power amplifier section, was able to demonstrate a peak power of 1.2 kW for 105 ns pulses at 1530 nm, equivalent to a pulse energy of 0.126 mJ, with transform-limited linewidth and diffraction limited beam quality [91].

In 2012 a pulse energy of 0.38 mJ and a peak power of 128 kW were demonstrated using a multiple stage, short, large core Yb/Er co-doped phosphate fiber, in which a short pulse width of few nanoseconds was used to further enhance the SBS threshold [92]. In fact, SBS is an acousto-optic non-linear effect due to the interaction between photons and phonons in the fiber, therefore employing pulses with duration shorter then phonon lifetime, in applications in which this is allowed, is an additional method to increase the SBS threshold.

Further increase in power amplifier performances was reported in 2014 by NP Photonics [93]. An all-fiber pulsed MOPA system with 1 W average power, 200 μJ energy, >10 kW peak power and 20 ns of duration at 1550 nm was demonstrated using a seed phosphate single-frequency fiber laser, amplified by several Er-doped phosphate fibers, with a final power amplifier stage consisting in 11 cm of highly Yb/Er co-doped PM phosphate gain fiber.

The same group reported also a pulsed all-fiber MOPA system in the nanosecond regime emitting in the 1064 nm band. The architecture was similar to the one at 1550 nm, but in this case Yb-doped phosphate fibers were used. In particular, the power amplifier consisted in 25 cm of highly $Yb^{3+}$-doped PM phosphate fiber.

Both MOPA systems were packaged and shipped as an Original Equipment Manufacturer (OEM) product, just showing the maturity level achieved by phosphate fiber laser technology.

The main results obtained in the 1.5 μm region with high energy and peak power pulsed single-frequency all-fiber amplifiers are summarized in Table 4.

**Table 4.** Main results obtained in the 1.5 μm region with high energy, high peak power pulsed single-frequency all-fiber amplifiers based on highly $Yb^{3+}/Er^{3+}$ co-doped phosphate glasses.

| Reference | $\lambda_{signal}$ (nm) | Pulse Energy (mJ) | Pulse Duration (ns) | Average Power (W) | Beam Quality ($M^2$) | Peak Power (kW) |
|---|---|---|---|---|---|---|
| [21] | 1538 | 0.054 | 153 | 1.08 | 1.2 | 0.332 |
| [91] | 1530 | 0.126 | 100 | - | 1.2–1.4 | 1.2 |
| [92] | 1550 | 0.384 | 3 | - | - | 128 |
| [93] | 1550 | 0.200 | 20 | 1 | - | >10 |

## 6. Conclusions

In the last decades, remarkable achievements in the field of phosphate based fiber lasers and amplifiers have been obtained thanks to the progress in the diode technology, advanced fiber fabrication technology and novel pumping techniques. In this review, we reported on current state-of-the-art CW, pulsed, and single-frequency fiber lasers and amplifiers based on phosphate glass optical fibers. Phosphate glasses main properties were also briefly discussed to give to the readers a general overview of this outstanding gain medium.

The reported results show that phosphate glass fiber lasers and amplifiers can be highly efficient in a short fiber laser arrangement and can thus represent a promising alternative for other glass lasers at medium output power performance. Thus, in future, phosphate based fiber lasers and amplifiers are expected to be extensively employed in different applications fields.

**Author Contributions:** This article has been written by Nadia Giovanna Boetti and revised by Diego Pugliese, Edoardo Ceci-Ginistrelli, Joris Lousteau, Davide Janner and Daniel Milanese.

**Conflicts of Interest:** The authors declare no conflict of interest.

## References

1.  Snitzer, E. Optical maser action of $Nd^{+3}$ in a barium crown glass. *Phys. Rev. Lett.* **1961**, *7*, 444–446. [CrossRef]
2.  Koester, C.J.; Snitzer, E. Amplification in a fiber laser. *Appl. Opt.* **1964**, *7*, 1182–1186. [CrossRef]
3.  Richardson, D.J.; Nilsson, J.; Clarkson, W.A. High power fiber lasers: Current status and future perspectives. *J. Opt. Soc. Am. B* **2010**, *27*, B63–B92. [CrossRef]
4.  Nilsson, J.; Payne, D.N. High-power fiber lasers. *Science* **2011**, *332*, 921–922. [CrossRef] [PubMed]
5.  Richardson, K.; Krol, D.; Hirao, K. Glasses for photonic applications. *Int. J. Appl. Glass Sci.* **2010**, *1*, 74–86. [CrossRef]
6.  Jiang, S.; Mendes, S.B.; Hu, Y.; Nunzi Conti, G.; Chavez-Pirson, A.; Kaneda, Y.; Luo, T.; Chen, Q.; Hocde, S.; Nguyen, D.T.; et al. Compact multimode pumped erbium-doped phosphate fiber amplifiers. *Opt. Eng.* **2003**, *42*, 2817–2820. [CrossRef]
7.  Qiu, T.; Li, L.; Schülzgen, A.; Temyanko, V.L.; Luo, T.; Jiang, S.; Mafi, A.; Moloney, J.V.; Peyghambarian, N. Generation of 9.3-W multimode and 4-W single-mode output from 7-cm short fiber lasers. *IEEE Photonics Technol. Lett.* **2004**, *16*, 2592–2594. [CrossRef]
8.  Lee, Y.-W.; Sinha, S.; Digonnet, M.J.F.; Byer, R.L.; Jiang, S. 20 W single-mode $Yb^{3+}$-doped phosphate fiber laser. *Opt. Lett.* **2006**, *31*, 3255–3257. [CrossRef] [PubMed]
9.  Izumitani, T.S. *Optical Glass*, 3rd ed.; American Institute of Physics: College Park, MD, USA, 1993; ISBN 0883185067.
10. Seneschal, K.; Smektala, F.; Bureau, B.; Le Floch, M.; Jiang, S.; Luo, T.; Lucas, J.; Peyghambarian, N. Properties and structure of high erbium doped phosphate glass for short optical fibers amplifiers. *Mater. Res. Bull.* **2005**, *40*, 1433–1442. [CrossRef]
11. Hwang, B.-C.; Jiang, S.; Luo, T.; Watson, J.; Sorbello, G.; Peyghambarian, N. Cooperative upconversion and energy transfer of new high $Er^{3+}$- and $Yb^{3+}$–$Er^{3+}$-doped phosphate glasses. *J. Opt. Soc. Am. B* **2000**, *17*, 833–839. [CrossRef]
12. Pugliese, D.; Boetti, N.G.; Lousteau, J.; Ceci-Ginistrelli, E.; Bertone, E.; Geobaldo, F.; Milanese, D. Concentration quenching in an Er-doped phosphate glass for compact optical lasers and amplifiers. *J. Alloys Compd.* **2016**, *657*, 678–683. [CrossRef]
13. Ohtsuki, T.; Honkanen, S.; Najafi, S.I.; Peyghambarian, N. Cooperative upconversion effects on the performance of $Er^{3+}$-doped phosphate glass waveguide amplifiers. *J. Opt. Soc. Am. B* **1997**, *14*, 1838–1845. [CrossRef]
14. Jiang, C.; Hu, W.; Zeng, Q. Numerical analysis of concentration quenching model of $Er^{3+}$-doped phosphate fiber amplifier. *IEEE J. Quantum Electron.* **2003**, *39*, 1266–1271. [CrossRef]
15. Lee, Y.W.; Sinha, S.; Digonnet, M.J.F.; Byer, R.L.; Jiang, S. Measurement of high photodarkening resistance in heavily $Yb^{3+}$-doped phosphate fibres. *Electron. Lett.* **2008**, *44*, 14–16. [CrossRef]
16. Campbell, J.H.; Hayden, J.S.; Marker, A. High-power solid-state lasers: A laser glass perspective. *Int. J. Appl. Glass Sci.* **2011**, *2*, 3–29. [CrossRef]
17. Boetti, N.G.; Lousteau, J.; Mura, E.; Abrate, S.; Milanese, D. CW cladding pumped phosphate glass fibre laser operating at 1.054 μm. In Proceedings of the 16th International Conference on Transparent Optical Networks (ICTON), Graz, Austria, 6–10 July 2014; IEEE: New York, NY, USA, 2014. [CrossRef]
18. SciGlass-Glass Property Information System. Available online: http://www.akosgmbh.de/sciglass/sciglass.htm (accessed on 25 October 2017).
19. Jiang, S. Erbium-doped phosphate fiber amplifiers. In Proceedings of the SPIE 5246, Active and Passive Optical Components for WDM Communications III, Information Technologies and Communications (ITCom 2003), Orlando, FL, USA, 7–11 September 2003; SPIE: Bellingham, DC, USA, 2003. [CrossRef]

20. Zhu, X.; Shi, W.; Zong, J.; Nguyen, D.; Norwood, R.A.; Chavez-Pirson, A.; Peyghambarian, N. 976 nm single-frequency distributed Bragg reflector fiber laser. *Opt. Lett.* **2012**, *37*, 4167–4169. [CrossRef] [PubMed]
21. Shi, W.; Petersen, E.B.; Leigh, M.; Zong, J.; Yao, Z.; Chavez-Pirson, A.; Peyghambarian, N. High SBS-threshold single-mode single-frequency monolithic pulsed fiber laser in the C-band. *Opt. Express* **2009**, *17*, 8237–8245. [CrossRef] [PubMed]
22. Zervas, M.N.; Codemard, C.A. High power fiber lasers: A review. *IEEE J. Sel. Top. Quantum Electron.* **2014**, *20*, 1–23. [CrossRef]
23. Pask, H.M.; Carman, R.J.; Hanna, D.C.; Tropper, A.C.; Mackechnie, C.J.; Barber, P.R.; Dawes, J.M. Ytterbium-doped silica fiber lasers: Versatile sources for the 1–12 µm region. *IEEE J. Sel. Top. Quantum Electron.* **1995**, *1*, 2–13. [CrossRef]
24. Digonnet, M.J.F. *Rare-Earth-Doped Fiber Lasers and Amplifiers*, 2nd ed.; Marcel Dekker, Inc.: New York, NY, USA, 2001; ISBN 0-8247.
25. Boetti, N.G.; Negro, D.; Lousteau, J.; Freyria, F.S.; Bonelli, B.; Abrate, S.; Milanese, D. Spectroscopic investigation of $Nd^{3+}$ single doped and $Eu^{3+}/Nd^{3+}$ co-doped phosphate glass for solar pumped lasers. *J. Non-Cryst. Solids* **2013**, *377*, 100–104. [CrossRef]
26. Boetti, N.G.; Lousteau, J.; Negro, D.; Mura, E.; Scarpignato, G.C.; Perrone, G.; Abrate, S. Solar pumping of solid state lasers for space mission: A novel approach. In Proceedings of the International Conference on Space Optics 2012, Ajaccio, France, 9–12 October 2012; SPIE: Bellingham, WA, USA, 2012.
27. Ceci-Ginistrelli, E.; Smith, C.; Pugliese, D.; Lousteau, J.; Boetti, N.G.; Clarkson, W.A.; Poletti, F.; Milanese, D. Nd-doped phosphate glass cane laser: From materials fabrication to power scaling tests. *J. Alloys Compd.* **2017**, *722*, 599–605. [CrossRef]
28. Dubinskii, M.; Zhang, J.; Ter-Mikirtychev, V. Highly scalable, resonantly cladding-pumped, Er-doped fiber laser with record efficiency. *Opt. Lett.* **2009**, *34*, 1507–1509. [CrossRef] [PubMed]
29. Zhang, J.; Fromzel, V.; Dubinskii, M. Resonantly cladding-pumped Yb-free Er-doped LMA fiber laser with record high power and efficiency. *Opt. Express* **2011**, *19*, 5574–5578. [CrossRef] [PubMed]
30. Gapontsev, V.P.; Matitsin, S.M.; Isineev, A.A.; Kravchenko, V.B. Erbium glass lasers and their applications. *Opt. Laser Technol.* **1982**, *14*, 189–196. [CrossRef]
31. Reisfeld, R.; Jørgensen, C.K. Excited state phenomena in vitreous materials. In *Handbook on the Physics and Chemistry of Rare Earths*, 1st ed.; Gschneidner, K.A., Eyring, L., Eds.; Elsevier Science Publishers B.V.: North-Holland, The Netherlands, 1987; Volume 9, pp. 1–99, ISBN 978-0444820143.
32. Bufetov, I.A.; Semenov, S.L.; Kosolapov, A.F.; Mel'kumov, M.A.; Dudin, V.V.; Galagan, B.I.; Denker, B.I.; Osiko, V.V.; Sverchkov, S.E.; Dianov, E.M. Ytterbium fibre laser with a heavily $Yb^{3+}$-doped glass fibre core. *Quantum Electron.* **2006**, *36*, 189–191. [CrossRef]
33. Payne, S.A.; Marshall, C.D.; Bayramian, A.; Wilke, G.D.; Hayden, J.S. Laser properties of a new average-power Nd-doped phosphate glass. *Appl. Phys. B* **1995**, *61*, 257–266. [CrossRef]
34. Lafond, C.; Osouf, J.; Laperle, P.; Soucy, J.-L.; Desrosiers, C.; Morency, S.; Croteau, A.; Parent, A. $Er^{3+}$-$Yb^{3+}$ co-doped phosphate glass optical fiber for application at 1.54 microns. In Proceedings of the SPIE 6343, Photonics North 2006, Quebec City, QC, Canada, 19–22 June 2006; SPIE: Bellingham, WA, USA, 2006. [CrossRef]
35. Nguyen, D.T.; Chavez-Pirson, A.; Jiang, S.; Peyghambarian, N. A novel approach of modeling cladding-pumped highly Er–Yb co-doped fiber amplifiers. *IEEE J. Quantum Electron.* **2007**, *43*, 1018–1027. [CrossRef]
36. Snitzer, E.; Po, H.; Hakimi, F.; Tumminelli, R.; McCollum, B.C. Double clad, offset core Nd fiber laser. In Proceedings of the Optical Fiber Sensors 1988, New Orleans, LA, USA, 27 January 1988; OSA: Washington, DC, USA, 1988. [CrossRef]
37. IPG Set to Ship 100 kW Laser. Available online: http://optics.org/news/3/10/44 (accessed on 25 October 2017).
38. Stiles, E. New developments in IPG fiber laser technology. In Proceedings of the 5th International Workshop on Fiber Lasers, Dresden, Germany, 30 September–1 October 2009.
39. Jeong, Y.; Yoo, S.; Codemard, C.A.; Nilsson, J.; Sahu, J.K.; Payne, D.N.; Horley, R.; Turner, P.W.; Hickey, L.; Harker, A.; et al. Erbium:Ytterbium codoped large-core fiber laser with 297-W continuous-wave output power. *IEEE J. Sel. Top. Quantum Electron.* **2007**, *13*, 573–579. [CrossRef]

40. Ehrenreich, T.; Leveille, R.; Majid, I.; Tankala, K.; Rines, G.; Moulton, P. 1-kW, all-glass Tm:fiber laser. In Proceedings of the Fiber Lasers VII: Technology, Systems, and Applications, San Francisco, CA, USA, 25–28 January 2010; SPIE: Bellingham, DC, USA, 2010. [CrossRef]

41. Shen, X.; Zhang, L.; Ding, J.; Wei, W. Design, fabrication, and optical gain performance of the gain-guided and index-antiguided $Nd^{3+}$-doped phosphate glass fiber. *J. Opt. Soc. Am. B* **2017**, *34*, 998–1003. [CrossRef]

42. Wang, L.; He, D.; Hu, L.; Chen, D. $Nd^{3+}$-doped soft glass double-clad fibers with a hexagonal inner cladding. *Laser Phys.* **2015**, *25*, 045108. [CrossRef]

43. Lee, Y.-W.; Digonnet, M.J.F.; Sinha, S.; Urbanek, K.E.; Byer, R.L.; Jiang, S. High-power $Yb^{3+}$-doped phosphate fiber amplifier. *IEEE J. Sel. Top. Quantum Electron.* **2009**, *15*, 93–102. [CrossRef]

44. Franczyk, M.; Stępień, R.; Piechal, B.; Pysz, D.; Stawicki, K.; Siwicki, B.; Buczyński, R. High efficiency $Yb^{3+}$-doped phosphate single-mode fibre laser. *Laser Phys. Lett.* **2017**, *14*, 105102. [CrossRef]

45. Yamashita, T. Nd- and Er-doped phosphate glass for fiber laser. In Proceedings of the SPIE 1171, Fiber Laser Sources and Amplifiers, OE/FIBERS'89, Boston, MA, USA, 5–8 September 1989; SPIE: Bellingham, WA, USA, 1989. [CrossRef]

46. Griebner, U.; Koch, R.; Schönnagel, H.; Grunwald, R. Efficient laser operation with nearly diffraction-limited output from a diode-pumped heavily Nd-doped multimode fiber. *Opt. Lett.* **1996**, *21*, 266–268. [CrossRef] [PubMed]

47. Zhang, G.; Wang, M.; Yu, C.; Zhou, Q.; Qiu, J.; Hu, L.; Chen, D. Efficient generation of watt-level output from short-length Nd-doped phosphate fiber lasers. *IEEE Photonics Technol. Lett.* **2011**, *23*, 350–352. [CrossRef]

48. Zhang, G.; Yu, C.L.; Wang, M.; Zhou, Q.L.; Qiu, J.R.; Hu, L.L.; Chen, D.P. Local-cooled watt-level Nd-doped phosphate single-mode fiber laser. *Laser Phys.* **2012**, *22*, 1235–1239. [CrossRef]

49. Délen, X.; Piehler, S.; Didierjean, J.; Aubry, N.; Voss, A.; Ahmed, M.A.; Graf, T.; Balembois, F.; Georges, P. 250 W single-crystal fiber Yb: YAG laser. *Opt. Lett.* **2012**, *37*, 2898–2900. [CrossRef] [PubMed]

50. Zhang, G.; Zhou, Q.; Yu, C.; Hu, L.; Chen, D. Neodymium-doped phosphate fiber lasers with an all-solid microstructured inner cladding. *Opt. Lett.* **2012**, *37*, 2259–2261. [CrossRef] [PubMed]

51. Wang, L.; Liu, H.; He, D.; Yu, C.; Hu, L.; Qiu, J.; Chen, D. Phosphate single mode large mode area all-solid photonic crystal fiber with multi-watt output power. *Appl. Phys. Lett.* **2014**, *104*, 131111. [CrossRef]

52. Wang, L.; He, D.; Feng, S.; Yu, C.; Hu, L.; Chen, D. Seven-core Neodymium-doped phosphate all-solid photonic crystal fibers. *Laser Phys.* **2016**, *26*, 015104. [CrossRef]

53. Jiang, S.; Hwang, B.-C.; Luo, T.; Seneschal, K.; Peyghambarian, N.; Smektala, F.; Honkanen, S.; Lucas, J. Net gain of 15.5 dB from a 5.1 cm-long $Er^{3+}$-doped phosphate glass fiber. In Proceedings of the Optical Fiber Communication (OFC) Conference 2000, Baltimore, MD, USA, 7–10 March 2000; OSA: Washington, DC, USA, 2000.

54. Hwang, B.-C.; Jiang, S.; Luo, T.; Seneschal, K.; Sorbello, G.; Morrell, M.; Smektala, F.; Honkanen, S.; Lucas, J.; Peyghambarian, N. Performance of high-concentration $Er^{3+}$-doped phosphate fiber amplifiers. *IEEE Photonics Technol. Lett.* **2001**, *13*, 197–199. [CrossRef]

55. Hu, Y.; Jiang, S.; Luo, T.; Seneschal, K.; Morrell, M.; Smektala, F.; Honkanen, S.; Lucas, J.; Peyghambarian, N. Performance of high-concentration $Er^{3+}$-$Yb^{3+}$-codoped phosphate fiber amplifiers. *IEEE Photonics Technol. Lett.* **2001**, *13*, 657–659. [CrossRef]

56. Xu, S.H.; Yang, Z.M.; Feng, Z.M.; Zhang, Q.Y.; Jiang, Z.H.; Xu, W.C. Gain and noise characteristics of single-mode $Er^{3+}$/$Yb^{3+}$ co-doped phosphate glass fibers. In Proceedings of the 2nd IEEE Nanoelectronics Conference (INEC) 2008, Shanghai, China, 24–27 March 2008; IEEE: New York, NY, USA, 2008. [CrossRef]

57. Boetti, N.G.; Scarpignato, G.C.; Lousteau, J.; Pugliese, D.; Bastard, L.; Broquin, J.-E.; Milanese, D. High concentration Yb-Er co-doped phosphate glass for optical fiber amplification. *J. Opt.* **2015**, *17*, 065705. [CrossRef]

58. Kuhn, V.; Kracht, D.; Neumann, J.; Weßels, P. Er-doped single-frequency photonic crystal fiber amplifier with 70 W of output power for gravitational wave detection. In Proceedings of the SPIE 8237, Fiber Lasers IX: Technology, Systems, and Application (SPIE LASE 2012), San Francisco, CA, USA, 21–26 January 2012; SPIE: Bellingham, WA, USA, 2012. [CrossRef]

59. Canat, G.; Augère, B.; Besson, C.; Dolfi-Bouteyre, A.; Durecu, A.; Goular, D.; Le Gouët, J.; Lombard, L.; Planchat, C.; Valla, M. High peak power single-frequency MOPFA for Lidar applications. In Proceedings of the Conference on Lasers and Electro-Optics (CLEO), San Jose, CA, USA, 5–10 June 2016; OSA: Washington, DC, USA, 2016. [CrossRef]

60. Shi, W.; Leigh, M.A.; Zong, J.; Yao, Z.; Nguyen, D.T.; Chavez-Pirson, A.; Peyghambarian, N. High-power all-fiber-based narrow-linewidth single-mode fiber laser pulses in the C-band and frequency conversion to THz generation. *IEEE J. Sel. Top. Quantum Electron.* **2009**, *15*, 377–384. [CrossRef]

61. Leger, J.R.; Nilsson, J.; Huignard, J.P.; Napartovich, A.P.; Shay, T.M.; Shirakawa, A. Introduction to the issue on laser beam combining and fiber laser systems. *IEEE J. Sel. Top. Quantum Electron.* **2009**, *15*, 237–239. [CrossRef]

62. Claps, R.; Sabbaghzadeh, J.; Fink, M. Raman spectroscopy with a single-frequency, high power, broad-area laser diode. *Appl. Spectrosc.* **1999**, *53*, 491–496. [CrossRef]

63. Fu, S.; Shi, W.; Feng, Y.; Zhang, L.; Yang, Z.; Xu, S.; Zhu, X.; Norwood, R.A.; Peyghambarian, N. Review of recent progress on single-frequency fiber lasers. *J. Opt. Soc. Am. B* **2017**, *34*, A49–A62. [CrossRef]

64. Balliu, E.; Engholm, M.; Hellström, J.; Elgcrona, G.; Karlsson, H. Compact nanosecond pulsed single stage Yb-doped fiber amplifier. In Proceedings of the SPIE 8959, Solid State Lasers XXIII: Technology and Devices (SPIE LASE 2014), San Francisco, CA, USA, 1–6 February 2014; SPIE: Bellingham, DC, USA, 2014. [CrossRef]

65. Mermelstein, M.D.; Andrejco, M.J.; Fini, J.; Yablon, A.; Headley, C.; Di Giovanni, D.J.; McCurdy, A.H. 11.2 dB SBS gain suppression in a large mode area Yb-doped optical fiber. In Proceedings of the SPIE 6873, Fiber Lasers V: Technology, Systems, and Application, Lasers and Applications in Science and Engineering (SPIE LASE 2008), San Jose, CA, USA, 22–24 January 2008; SPIE: Bellingham, DC, USA, 2008. [CrossRef]

66. Liu, A. Suppressing stimulated Brillouin scattering in fiber amplifiers using nonuniform fiber and temperature gradient. *Opt. Express* **2007**, *15*, 977–984. [CrossRef] [PubMed]

67. Products. Available online: http://www.npphotonics.com/ (accessed on 25 October 2017).

68. Kaneda, Y.; Spiegelberg, C.; Geng, J.; Hu, Y.; Luo, T.; Wang, J.; Jiang, S. 200-mW, narrow-linewidth 1064.2-nm Yb-doped fiber laser. In Proceedings of the Conference on Lasers and Electro-Optics (CLEO) 2004, San Francisco, CA, USA, 16–21 May 2004; OSA: Washington, DC, USA, 2004. [CrossRef]

69. Xu, S.; Yang, Z.; Zhang, W.; Wei, X.; Qian, Q.; Chen, D.; Zhang, Q.; Shen, S.; Peng, M.; Qiu, J. 400 mW ultrashort cavity low-noise single-frequency $Yb^{3+}$-doped phosphate fiber laser. *Opt. Lett.* **2011**, *36*, 3708–3710. [CrossRef] [PubMed]

70. Xu, S.; Li, C.; Zhang, W.; Mo, S.; Yang, C.; Wei, X.; Feng, Z.; Qian, Q.; Shen, S.; Peng, M.; et al. Low noise single-frequency single-polarization Ytterbium-doped phosphate fiber laser at 1083 nm. *Opt. Lett.* **2013**, *38*, 501–503. [CrossRef] [PubMed]

71. Yang, C.; Zhao, Q.; Feng, Z.; Peng, M.; Yang, Z.; Xu, S. 1120 nm kHz-linewidth single-polarization single-frequency Yb-doped phosphate fiber laser. *Opt. Express* **2016**, *24*, 29794–29799. [CrossRef] [PubMed]

72. Wu, J.; Zhu, X.; Temyanko, V.; LaComb, L.; Norwood, R.; Peyghambarian, N. Power scaling of single-frequency fiber amplifiers at 976 nm. In Proceedings of the Conference on Lasers and Electro-Optics (CLEO), San Jose, CA, USA, 5–10 June 2016; OSA: Washington, DC, USA, 2016. [CrossRef]

73. Wu, J.; Zhu, X.; Temyanko, V.; LaComb, L.; Kotov, L.; Kiersma, K.; Zong, J.; Li, M.; Chavez-Pirson, A.; Norwood, R.A.; et al. $Yb^{3+}$-doped double-clad phosphate fiber for 976 nm single-frequency laser amplifiers. *Opt. Mater. Express* **2017**, *7*, 1310–1316. [CrossRef]

74. Zhu, X.; Zhu, G.; Shi, W.; Zong, J.; Wiersma, K.; Nguyen, D.; Norwood, R.A.; Chavez-Pirson, A.; Peyghambarian, N. 976 nm Single-polarization single-frequency Ytterbium-doped phosphate fiber amplifiers. *IEEE Photonics Technol. Lett.* **2013**, *25*, 1365–1368. [CrossRef]

75. Mo, S.; Xu, S.; Huang, X.; Zhang, W.; Feng, Z.; Chen, D.; Yang, T.; Yang, Z. A 1014 nm linearly polarized low noise narrow-linewidth single-frequency fiber laser. *Opt. Express* **2013**, *21*, 12419–12423. [CrossRef] [PubMed]

76. Yang, C.; Xu, S.; Yang, Q.; Lin, W.; Mo, S.; Li, C.; Feng, Z.; Chen, D.; Yang, Z.; Jiang, Z. High-efficiency watt-level 1014 nm single-frequency laser based on short Yb-doped phosphate fiber amplifiers. *Appl. Phys. Express* **2014**, *7*, 062702. [CrossRef]

77. Spiegelberg, C.; Geng, J.; Hu, Y.; Luo, T.; Kaneda, Y.; Wang, J.; Li, W.; Brutsch, M.; Hocde, S.; Chen, M.; et al. Compact 100 mW fiber laser with 2 kHz linewidth. In Proceedings of the Optical Fiber Communications Conference (OFC), Atlanta, GA, USA, 23–28 March 2003; IEEE: New York, NY, USA, 2003. [CrossRef]

78. Hofmann, P.; Voigtländer, C.; Nolte, S.; Peyghambarian, N.; Schülzgen, A. 550-mW output power from a narrow linewidth all-phosphate fiber laser. *J. Lightwave Technol.* **2013**, *31*, 756–760. [CrossRef]

79. Qiu, T.; Suzuki, S.; Schülzgen, A.; Li, L.; Polynkin, A.; Temyanko, V.; Moloney, J.V.; Peyghambarian, N. Generation of watt-level single-longitudinal-mode output from cladding-pumped short fiber lasers. *Opt. Lett.* **2005**, *30*, 2748–2750. [CrossRef] [PubMed]

80. Morkel, P.R.; Jedrzejewski, K.P.; Taylor, E.R. Q-switched Neodymium-doped phosphate glass fiber lasers. *IEEE J. Sel. Top. Quantum Electron.* **1993**, *29*, 2178–2188. [CrossRef]

81. Kaneda, Y.; Hu, Y.; Spiegelberg, C.; Geng, J.; Jiang, S. Single-frequency, all-fiber Q-switched laser at 1550-nm. In Proceedings of the Advanced Solid-State Photonics, Santa Fe, NM, USA, 1–4 February 2004; OSA: Washington, DC, USA, 2004. [CrossRef]

82. Leigh, M.; Shi, W.; Zong, J.; Wang, J.; Jiang, S.; Peyghambarian, N. Compact, single-frequency all-fiber Q-switched laser at 1 μm. *Opt. Lett.* **2007**, *32*, 897–899. [CrossRef] [PubMed]

83. Yamashita, S.; Yoshida, T.; Set, S.Y.; Polynkin, P.; Peyghambarian, N. Passively mode-locked short-cavity 10 GHz Er:Yb-codoped phosphate-fiber laser using carbon nanotubes. In Proceedings of the SPIE 6453, Fiber Lasers IV: Technology, Systems, and Applications (SPIE LASE 2007), San Jose, CA, USA, 20–25 January 2007; SPIE: Bellingham, DC, USA, 2007. [CrossRef]

84. Ye, N.N.; Pan, Z.Q.; Yang, F.; Ye, Q.; Cai, H.W.; Qu, R.H. 7-GHz high-repetition-rate mode-locked pulse generation using short-cavity phosphate glass fiber laser. *Laser Phys.* **2012**, *22*, 1247–1251. [CrossRef]

85. Thapa, R.; Nguyen, D.; Zong, J.; Chavez-Pirson, A. All-fiber fundamentally mode-locked 12 GHz laser oscillator based on an Er/Yb-doped phosphate glass fiber. *Opt. Lett.* **2014**, *39*, 1418–1421. [CrossRef] [PubMed]

86. Shi, W.; Fang, Q.; Zhu, X.; Norwood, R.A.; Peyghambarian, N. Fiber lasers and their applications. *Appl. Opt.* **2014**, *53*, 6554–6568. [CrossRef] [PubMed]

87. Agrawal, G.P. *Nonlinear Fiber Optics*, 5th ed.; Academic Press: Cambridge, MA, USA, 2012; ISBN 9780123970237.

88. Kovalev, V.I.; Harrison, R.G. Suppression of stimulated Brillouin scattering in high-power single-frequency fiber amplifiers. *Opt. Lett.* **2006**, *31*, 161–163. [CrossRef] [PubMed]

89. Yoshizawa, N.; Imai, T. Stimulated Brillouin scattering suppression by means of applying strain distribution to fiber with cabling. *J. Lightwave Technol.* **1993**, *11*, 1518–1522. [CrossRef]

90. Chavez Boggio, J.M.; Marconi, J.D.; Fragnito, H.L. Experimental and numerical investigation of the SBS-threshold increase in an optical fiber by applying strain distributions. *J. Lightwave Technol.* **2005**, *23*, 3808–3814. [CrossRef]

91. Shi, W.; Petersen, E.B.; Yao, Z.; Nguyen, D.T.; Zong, J.; Stephen, M.A.; Chavez-Pirson, A.; Peyghambarian, N. Kilowatt-level stimulated-Brillouin-scattering-threshold monolithic transform-limited 100 ns pulsed fiber laser at 1530 nm. *Opt. Lett.* **2010**, *35*, 2418–2420. [CrossRef] [PubMed]

92. Petersen, E.; Shi, W.; Chavez-Pirson, A.; Peyghambarian, N. High peak-power single-frequency pulses using multiple stage, large core phosphate fibers and preshaped pulses. *Appl. Opt.* **2012**, *51*, 531–534. [CrossRef] [PubMed]

93. Akbulut, M.; Miller, A.; Wiersma, K.; Zong, J.; Rhonehouse, D.; Nguyen, D.T.; Chavez-Pirson, A. High energy, high average and peak power phosphate-glass fiber amplifiers for 1 micron band. In Proceedings of the SPIE 8961, Fiber Lasers XI: Technology, Systems, and Applications (SPIE LASE 2014), San Francisco, CA, USA, 1–6 February 2014; SPIE: Bellingham, WA, USA, 2014. [CrossRef]

*applied*
*sciences*

MDPI

*Review*

# High-Power, Solid-State, Deep Ultraviolet Laser Generation

Hongwen Xuan [1,*], Hironori Igarashi [2], Shinji Ito [1], Chen Qu [2], Zhigang Zhao [1] and Yohei Kobayashi [1]

[1]   The Institute for Solid State Physics, the University of Tokyo, Kashiwanoha 5-1-5, Kashiwa, Chiba 277-8581, Japan; shinji_ito@issp.u-tokyo.ac.jp (S.I.); zhigang@issp.u-tokyo.ac.jp (Z.Z.); yohei@issp.u-tokyo.ac.jp (Y.K.)
[2]   GIGAPHOTON INC., 400 Yokokurashinden, Oyama, Tochigi 323-8558, Japan; hironori_igarashi@gigaphoton.com (H.I.); chen_qu@gigaphoton.com (C.Q.)
*    Correspondence: xuan@issp.u-tokyo.ac.jp; Tel.: +81-47-136-3356

Received: 8 November 2017; Accepted: 30 January 2018; Published: 3 February 2018

**Abstract:** At present, deep ultraviolet (DUV) lasers at the wavelength of fourth harmonics of 1 μm (266 nm/258 nm) and at the wavelength of 193 nm are widely utilized in science and industry. We review the generation of these DUV lasers by nonlinear frequency conversion processes using solid-state/fiber lasers as the fundamental frequency. A DUV laser at 258 nm by fourth harmonics generation (FHG) could achieve an average power of 10 W with a beam quality of $M^2 < 1.5$. Moreover, 1 W of average power at 193 nm was obtained by sum-frequency generation (SFG). A new concept of 193-nm DUV laser generation by use of the diamond Raman laser is also introduced. A proof-of-principle experiment of the diamond Raman laser is reported with the conversion efficiency of 23% from the pump to the second Stokes wavelength, which implies the potential to generate a higher power 193 nm DUV laser in the future.

**Keywords:** deep ultraviolet laser; frequency conversion; Raman laser

## 1. Introduction

Deep ultraviolet (DUV) lasers are currently widely employed in various applications. For instance, a DUV laser at 260 nm has been applied as an "external seed" of a free-electron laser (FEL) with output wavelengths as short as 4.3 nm, which would make it possible to do the scientific research beyond the carbon K-edge [1]. In the industrial applications, laser machining of wide bandgap materials would be also beneficial from the DUV lasers owing to their characteristics of high photon energy [2].

In the past decades, with the fast development of the solid-state lasers and the nonlinear optical crystals, high-power DUV lasers at 266 nm/258 nm have been studied by many research groups fruitfully [3–17]. These DUV lasers are usually generated by fourth-harmonics generation (FHG) of the near-infrared (NIR) laser at 1 μm as shown in Figure 1. Until now, the highest average output power of these DUV lasers was 40 W at the wavelength of 266 nm, of which the fundamental light was a high-power $Nd:Y_3Al_5O_{12}(Nd:YAG)$ laser at 1064 nm and the nonlinear optical crystal was a $CsLiB_6O_{10}$ (CLBO) [8]. In the scientific and industrial applications, the beam quality of the DUV laser is also an important parameter. Particularly, the beam quality of the DUV laser dramatically affects the laser machining results. The beam quality of the DUV laser is determined by that of the fundamental laser and the crystal for the frequency conversion. Recently, a high-power DUV laser (>10 W) at 258 nm was reported with a beam quality of $M^2 < 1.5$, which was generated by FHG of a Yb:YAG laser at 1030 nm using the CLBO crystal [13]. This laser would lead to a better laser machining result compared to the previous DUV lasers with larger $M^2$.

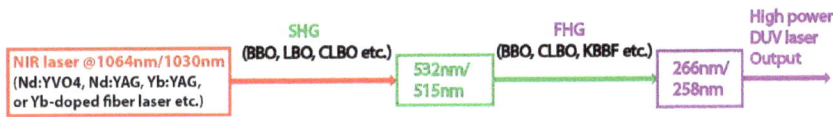

**Figure 1.** Schematic setup of the solid-state DUV laser by FHG of 1-μm laser (DUV: deep ultra-violet; SHG: second-harmonics generation; FHG: fourth-harmonics generation; YAG:$Y_3Al_5O_{12}$; BBO: $\beta$-$BaB_2O_4$; LBO: $LiB_3O_5$; CLBO: $CsLiB_6O_{10}$; KBBF: $KBe_2BO_3F_2$).

In terms of the solid-state DUV laser at 193 nm, the laser machining as well as other applications such as mask-inspection or lithography require a high-power DUV laser output [4,18,19]. The highest output power of the continuous-wave (CW) 193-nm laser was 120 mW by stages of sum-frequency generation (SFG) between DUV lasers (wavelength longer than 193 nm) and corresponding NIR lasers [4]. For the pulsed lasers, a DUV laser at 193 nm with average power of 300 mW was reported and had already been applied to be the seed of a hybrid ArF excimer laser [2,12]. By use of this hybrid ArF excimer laser schematics, both high power (>100 W) and high coherence could be obtained at the same time, which consists of a solid-state DUV laser seed and an ArF excimer amplifier [2]. A high-power solid-state DUV seeding laser with high coherence is beneficial to the final output of the hybrid ArF excimer laser, and it would also result in suppressing the amplified spontaneous emission (ASE) of the amplifier.

SFG is the most popular method to generate the DUV laser at 193 nm. Typically, a DUV laser, of which the wavelength is usually from 200 nm to 300 nm, and a NIR laser play the roles of SFG lasers in the nonlinear crystals to generate the 193-nm DUV laser. So far, the highest output power at 193 nm by following the above SFG schematics was 1 W by two stages of SFG [20]. As aforementioned, there is still great demand for power scaling of the solid-state DUV laser at 193 nm. Thus, by the SFG schematics, the requirement is to increase both the DUV laser (wavelength longer than 193 nm) power and the NIR laser power at the same time. By choosing the appropriate high-power fundamental laser at 1 μm as well as the nonlinear optical crystals, the average power of the DUV laser at the fourth harmonics of 1 μm could be obtained to be at the 10-watts level in the previous reports. On the other side, the power scaling of the NIR laser is another aspect to increase the average power of the 193-nm DUV laser. Due to the lack of the high-power amplification solution of the NIR laser (for example at 1553 nm), nonlinear frequency processes such as optical parameter oscillation (OPO) could be applied to build a wavelength converter by shifting the wavelength from 1 μm to the desired wavelength as well as obtaining relative higher power [21]. Recently, a diamond Raman laser showed its ability in achieving a high power and a high conversion efficiency at the 1st Stokes of the 1-μm laser [22,23], which implies the possibility to be applied in the 193-nm DUV laser generation because of its high Raman gain and large frequency shift.

In this paper, we present a review on the current status of the high-power solid-state DUV lasers. The high-power DUV laser at 258 nm was generated by FHG from 1 μm. The development of the 193-nm laser was demonstrated by stages of SFG. In order to scale the DUV laser power effectively, a new concept of the 193-nm DUV laser generation is presented based on the diamond Raman laser.

## 2. DUV Laser Generated by FHG of the 1-μm NIR Laser

Figure 1 depicts the schematic setup of the solid-state DUV laser at the FHG wavelength (258 nm or 266 nm) of 1-μm laser, and Table 1 lists the typical results by following this setup including a result by SFG in the $LiB_3O_5$ (LBO) crystal. So far, the highest average output power of it was 40 W, which was reported by M. Nishioka et al. in the year 2003 [8] as the improvement result of their 20.5 W DUV laser in 2000 [9]. The fundamental laser was an Q-switched diode-pumped Nd:YAG green laser at the wavelength of 532 nm. The 40 W 266 nm laser was obtained by second-harmonics generation (SHG) in a CLBO crystal. Nevertheless, the beam quality of the DUV laser was not good enough because the green laser had a $M^2$ of 10 [8]. In 2006, G. Wang et al. reported a 28.4 W average power DUV laser at 266 nm by SHG of a 532 nm laser with $M^2$ = 6.5 [10]. Q. Liu et al. presented the result of their DUV laser with average output power of 14.8 W by FHG of a Nd:YVO$_4$ laser at the wavelength of 1064 nm [3]. The output power of the fundamental Nd:YVO$_4$ laser was 160 W with the beam quality of $M_x^2$ = 1.28 and $M_y^2$ = 1.21. After SHG, the beam quality of green laser was $M_x^2$ = 1.44 and $M_y^2$ = 1.28. However, the beam quality of DUV laser will be remarkably decayed due to the large walk-off effect of the β-BaB$_2$O$_4$ (BBO) crystal for FHG to 266 nm [6]. O. Novak et al. reported a 6-W DUV laser at 257.5 nm by FHG in a CLBO crystal which effectively improved the DUV laser beam quality due to the smaller walk-off effect of the CLBO crystal [11] comparing to the previous results using BBO crystals [3–6]. However, the estimated $M^2$ of it was still more than 1.5. From the above-reported results, the key point to obtain a high-power, high-beam quality DUV laser is to choose a high power fundamental laser at 1 μm with a good beam quality and proper nonlinear crystals with a smaller walk-off effect for SHG and FHG. A fundamental laser with a good beam quality is crucial for the final DUV beam quality as well as the nonlinear crystal with a smaller walk-off angle.

**Table 1.** Typical results of the solid-state DUV laser at 258 nm & 266 nm (NIR: near-infrared; DUV: deep ultra-violet; FHG: fourth-harmonics generation; BBO: β-BaB$_2$O$_4$; CLBO: CsLiB$_6$O$_{10}$; KBBF: KBe$_2$BO$_3$F$_2$; NSBBF: NaSr$_3$Be$_3$B$_3$O$_9$F$_4$; YAB: YAl$_3$ (BO$_3$)$_4$; SFG: sum-frequency generation).

| Wavelength (nm) | FHG Crystal | Power/Energy | Repetition Rate | NIR to FHG Efficiency (%) | $M^2$ (DUV) | Reference |
|---|---|---|---|---|---|---|
| 266 | BBO | 14.8 W | 100 kHz (~ns) | ~10 | >1.5 | [3] |
| 257 | BBO | 3.2 W | 30 kHz (~ns) | 14.5 | >1.1 | [4] |
| 257.7 | BBO | 2.74 mJ | 1 kHz (~ps) | 9.7 | <3 | [5] |
| 257 | BBO | 1.1 W | 14.5 kHz (~ns) | ~31 | | [6] |
| 258 | BBO | 4.6 W | 796 kHz (~fs) | ~5.8 | 1.3 × 1.6 | [7] |
| 266 | CLBO | 40 W | 7 kHz (~ns) | <10 | >10 | [8] |
| 266 | CLBO | 20.5 W | 10 kHz (~ns) | 14 | >10 | [9] |
| 266 | CLBO | 28.4 W | 10 kHz (~ns) | <22 | >6.24 | [10] |
| 257.5 | CLBO | 6 W | 100 kHz (~ps) | 10 | >1.5 | [11] |
| 258 | CLBO | 3 W | 6 kHz (~ns) | 35 | <1.5 | [12] |
| 258 | CLBO | 10.5 W | 10 kHz (~ns) | 31 | <1.5 | [13] |
| 266 | KBBF | 7.86 W | 10 kHz (~ns) | <10 | >2 | [14] |
| 266 | YAB | 5.05 W | 65 kHz (~ns) | 6.3 | | [15] |
| 266 | LBO(SFG) | 3.3 W | 1 MHz (~ns) | 14 | | [16] |
| 266 | NSBBF | 280 μJ | 10 Hz (~ps) | 36 | | [17] |

Due to the thermal lens effect in the solid-state lasers, if the heating management is insufficient, it is not easy to obtain the solid-state laser with a high power and a high beam quality simultaneously. Recently, with the development of the thin disk laser and single-crystal fiber (SCF) laser, a high-power laser at the fundamental wavelength (1 μm) had been realized with a good beam quality even with a TEM$_{00}$ mode beam [4,24], which would be good candidates for the fundamental of the DUV laser generation. Another candidate for the fundamental is the high-power fiber laser which could be achieved by coherent beam combination with $M^2$ value less than 1.1 on both axes of the beam. Hence, 4.6 W DUV laser at 258 nm was obtained by FHG of this fiber laser with an $M^2$ value of 1.3 × 1.6, which is the highest average power of the femtoseconds laser at this DUV wavelength so far [7]. Additionally, using the fiber laser as the fundamental, 100-W ultraviolet laser at 343 nm has also been obtained by third harmonics generation with $M^2$ < 1.4 [25]. Nevertheless, these high-power fiber lasers were achieved by the coherent beam combination requiring electronics systems to lock different amplifier channels, which increases the cost and the complexity.

Many kinds of crystals could be used for FHG such as BBO, CLBO, KBe$_2$BO$_3$F$_2$ (KBBF), YAl$_3$ (BO$_3$)$_4$ (YAB), NaSr$_3$Be$_3$B$_3$O$_9$F$_4$ (NSBBF) etc. as shown in Table 1. For FHG from 515 nm to 258 nm, the efficient nonlinear coefficient ($d_{eff}$) of the BBO crystal is the highest one among these crystals [15]. However, the BBO crystal also has the largest walk-off angle which will lead to a decline of the DUV laser beam quality comparing to use the crystal with a smaller walk-off angle such as CLBO. LBO is also an excellent nonlinear crystal which is always applied to the SHG from the NIR to the green. The DUV laser is usually realized by two stages of SFG in the LBO crystals since there is no phase-matching angle for LBO to convert the green laser at 515 nm or 532 nm to the DUV directly by SHG [16]. Recently, the NSBBF crystal has also shown its potential to obtain a high conversion efficiency from 532 nm to 266 nm. The walk-off angle and $d_{eff}$ of the NSBBF are almost similar to those of CLBO [17].

The conversion efficiency from fundamental NIR to DUV is from 5% to 36% according to the reported results in Table 1. Increasing the conversion efficiency will reduce the burden on the design and setup of the fundamental laser amplifier as well as the electric power consumption. A good beam quality of the fundamental laser will also contribute to the increasing of the conversion efficiency as well as the appropriate power intensity of the NIR and green laser in the SHG and FHG process, respectively. The highest conversion efficiencies from NIR to green and green to DUV laser by use of CLBO crystal are 70% and around 50%, respectively [12], which corresponds to the conversion efficiency of 35% from fundamental to DUV. In this report, the average output power of this DUV laser was 3 W which was limited by the fundamental NIR laser power. Recently, a DUV laser with 10.5 W average power output was reported, and the conversion efficiency was 31% from fundamental to DUV [13]. Its beam quality was the best for the 258 nm/266 nm DUV laser with the average power higher than 10 W, of which the $M^2$ is less than 1.5.

This high-power, high-beam quality DUV laser is also following the schematic setup as depicted in Figure 1. In detail, the fundamental laser was at 1030 nm which was from a master oscillator power amplifier (MOPA). The frequency conversion stages consisted of a noncritical phase-matching (NCPM) LBO crystal ($5 \times 5 \times 30$ mm³) for SHG from 1030 nm to 515 nm and a CLBO crystal for FHG from 515 nm to 258 nm. The seed laser of the MOPA was a distributed-feedback (DFB) laser diode (LD), and it was electrically modulated by a pulse generator with the repetition rate of 160 kHz and with a pulse duration of 10 ns. The repetition rate was lowered down to 10 kHz by an acousto-optic modulator (AOM). After 3 stages of fiber amplifiers, the signal was amplified to about 2 W. One stage of Yb:YAG ceramics thin rod (CTR) amplifier and one stage of Yb:YAG SCF amplifier (TARANIS, Fibercryst Inc., Décines-Charpieu, France) were used as the main amplifiers. The Yb:YAG CTR was in the dimension of $\Phi = 1$ mm $\times$ L $= 40$ mm with $Yb^{3+}$ ion doping of 1 at. % as well as the Yb:YAG SCF. The pump source of this stage was a 940 nm LD with 100 W output and the output power was 25 W by use of a double-pass configuration. The Yb:YAG SCF amplifier was pumped by a 200 W LD at the wavelength of 969 nm which was stabilized with a volume Bragg grating. The Yb:YAG SCF amplifier was a single pass configuration with 35 W output at 100 W pumping. The conversion efficiency of SHG from NIR to green and FHG from green to DUV was 73% and 42%, respectively, corresponding to a conversion efficiency of 31% from NIR to DUV. This was highest conversion efficiency for the DUV lasers with average powers of more than 10 W. Furthermore, the beam quality was near-diffraction-limited for both amplifiers (Yb:YAG CTR: $M_x^2 = 1.06$ and $M_y^2 = 1.03$; Yb:YAG SCF: $M_x^2 = 1.02$ and $M_y^2 = 1.04$). Hence, the DUV laser had an excellent beam quality owing to the good beam quality of the fundamental laser and the optimized frequency conversion stages. The measured value of $M^2$ was $M_x^2 = 1.02$ and $M_y^2 = 1.05$ for DUV laser at 3 W. Although the DUV beam became a little worse at 9–10 W ($M_x^2 = 1.37$ and $M_y^2 = 1.46$), it is still the best laser beam quality with $M^2 < 1.5$ for the DUV laser power more than 10 W [13].

The average power of the NIR laser still has the potential to be improved to tens of watts such as 70 W [26] if the double-pass configuration could be utilized for the Yb:YAG SCF amplifier stage. Hence, if the conversion efficiency of 31% from NIR to DUV could be still maintained at this higher power condition, the average output power of the DUV laser at 258 nm could be more than 20 W, which would be a remarkable power level for the DUV laser and lead to the unanticipated DUV laser machining results. Moreover, for high-power DUV laser, the UV damage of the nonlinear crystal, including the windows for CLBO crystal and the mirrors, is another important parameter for long term and stable operation. On the other hand, the beam quality of this higher power DUV laser would become worse due to more serious thermal and nonlinear effects. Hence, the frequency conversion processes should be carefully designed to mitigate the aforementioned effects. Thus, a high-power DUV laser at the FHG of 1 μm with a good beam quality could be obtained and will be beneficial for obtaining the shorter wavelength DUV lasers such as 193 nm by SFG.

## 3. DUV Laser at 193 nm

The reported typical results of the solid-state DUV laser at 193 nm in the past decades are listed in Table 2.

**Table 2.** Typical results of the solid-state DUV laser at 193 nm (CW: continuous-wave; KABO: $K_2Al_2B_2O_7$).

| Wavelength (nm) | Method | Crystal | Power/Engergy | Repetition Rate | Reference |
|---|---|---|---|---|---|
| 193 | SHG | KBBF | 15 mW | CW | [27] |
| 193 | SHG | KBBF | 2.23 mW | 8 kHz (~ns) | [28] |
| 193 | SHG | KBBF | 1.05 W | 1 kHz (~ps) | [29] |
| 193 | SHG | KBBF | 0.2 W | 6 kHz (~ns) | [30] |
| 193.4 | SFG (234.1 nm + 1110 nm) | CLBO | 120 mW | CW | [18] |
| 193 | SFG (221 nm + 1547 nm) | CLBO | 140 mW | 200 kHz (~ns) | [31] |
| 193.5 | SFG (221 nm + 1553 nm) | CLBO | 310 mW | 6 kHz (~ns) | [12] |
| 193 | SFG (226 nm + 1342 nm) | CLBO | 230 mW | 6 kHz (~ns) | [32] |
| 193 | SFG (221 nm + 1553 nm) | CLBO | 1.02 W | 10 kHz (~ns) | [20] |
| 193 | SFG (235.8 nm + 1064 nm) | KABO | 0.2 W | 10 kHz (~ns) | [33] |
| 193.4 | SFG (266 nm + 708.6 nm) | BBO | 2.5 mW | 5 kHz (~ns) | [19] |
| 193 | SFG (258 nm + 774 nm) | BBO | 2 μJ | 20 Hz (~fs) | [34] |
| 193 | SFG (213 nm + 2100 nm) | LBO | 770 μJ | 60 Hz (~ns) | [21] |

Two methods are usually applied to generate the solid-state DUV laser at the wavelength shorter than 200 nm. One is by SHG directly from a longer wavelength; the other one is by stages of SFG. T. Thao Tran et al. and Y. Yang et al., respectively, summarized the nonlinear optical materials for DUV generation from 5th to 7th harmonics of Nd:YAG laser at 1064 nm [35,36] by SHG and SFG including CLBO, KBBF, LBO, BBO, $K_2Al_2B_2O_7$ (KABO) etc. Recently, coherent radiation in the vacuum ultraviolet (VUV) region was generated with a wavelength as short as 121 nm in strontium tetra borate (SBO) by "random quasi-phase matching" (RQPM), which was the shortest wavelength generated by second-order nonlinear optical process in a solid-state material [37]. However, this SBO crystal does not fit for the high-power DUV laser generation due to the low efficiency of around $10^{-5}$.

Table 2 summarizes the typical results of the solid-state DUV laser at 193 nm by use of the different nonlinear crystals. Among the above crystals, the KBBF crystal is currently known as the only one, enabling high power and short wavelength directly by SHG [28–30,36,38,39]. For the application of seeding the hybrid ArF excimer laser, the DUV laser should be at the wavelength of 193 nm. In 2008, H. Zhang reported the result of 2.23 mW at 193 nm average power output directly by SHG in KBBF [28]. Later, Kanai et al. reported a 1.05 W 193-nm laser also by SHG in KBBF with the pulse duration of several hundred of picoseconds, of which the fundamental laser was from a Ti:Sapphire amplifier [29]. This is also the highest average power of 193-nm laser by SHG until now. In 2012, S. Ito et al. presented a 0.2 W 193-nm laser by SHG in KBBF with nanoseconds pulse which is suitable for seeding the ArF laser [30]. In addition, CW DUV laser at 193 nm was reported by M. Scholz et al. with the output power of 15 mW in KBBF by SHG with which the fundamental laser was from an external cavity laser [27].

The wavelength conversion of SHG to 193 nm by using KBBF was successfully demonstrated by the aforementioned groups. However, the KBBF should be packed in a prism coupling device (PCD) consisting of $CaF_2$ or $SiO_2$ prisms to achieve the phase-matching condition [14,29]. Moreover, KBBF is significantly difficult to be grown into single crystal because of the weak interaction of alkali metal and fluorine cations [35,36], which limit its potential to be grown into large dimension and to be applied in generating a higher power 193-nm laser for seeding the hybrid ArF laser. Hence, the SFG process in other crystals is the practical method for high-power 193-nm laser generation. The cut-off wavelength of DUV laser generated by SHG in BBO is limited to 205 nm by a phase matching condition although its absorption edge could reach down to 189 nm [40]. Nevertheless, the BBO crystal could be used in SFG process to generate the 193 nm laser. J.J. Jacob et al. reported a 193.4 nm laser by SFG between a 266 nm and a 708.6 nm laser with the output power 2.5 mW [19]. Even more than 20 years ago, femtoseconds 193 nm laser was generated in BBO by SFG between 258 nm and 774 nm [34]. Another drawback of BBO crystal is a large walk-off effect, which would deteriorate the SFG laser beam quality. LBO crystal could also be used for 193-nm laser generation if the wavelengths fulfilling the phase-matching condition. Hamilton et al. reported the 193-nm laser generation between 213 nm and 2100 nm in LBO by SFG with the energy of 770 µJ [21]. LBO has a smaller walk-off angle leading to a better beam quality. However, the efficient nonlinear coefficient is smaller than that of the BBO or CLBO crystal, which limits the DUV laser power scaling. KABO is also an excellent nonlinear optical crystal for DUV laser generation. Umemura et al. reported a 193 nm laser with the average output power of 0.2 W which could run stable for 50 hours although KABO has the smallest $d_{eff}$ comparing to BBO, LBO, and CLBO [33]. For the CLBO crystal, J. Sakuma reported a 120 mW DUV laser output at 193.4 nm by SFG between 234.1 nm and 1110 nm, which is the highest solid-state CW 193-nm laser until now [18]. Other results of 193 nm generation by SFG in CLBO also could reach to the average power from 100 mW to 300 mW [12,31,32]. CLBO has a larger $d_{eff}$ than LBO and a smaller walk-off angle than BBO, which means it could lead to a higher power DUV laser with better beam quality at the same time. Hence, the CLBO crystal is a suitable crystal for 193-nm laser generation. However, CLBO is more hygroscopic comparing to BBO or LBO crystals. Packaging the CLBO crystal into a cell, heating the crystal up to more than 150 °C, and purging it by use of noble gas is the solution to avoid these problems [41]. In addition, the previous CLBO crystal for FHG to DUV laser at 258 nm was also treated in the above conditions.

In terms of the experimental setup, the two laser wavelengths for SFG should be chosen carefully in order to satisfy the phase-matching condition of the nonlinear optical crystals. The Nd:YAG, Yb:YAG, Nd:YVO$_4$ and Yb-doped fiber lasers could achieve both high power and high beam quality. According to the published results for generating 193 nm, these NIR lasers were used for the fundamental of a DUV laser with the wavelength longer than 193 nm such as 258 nm (4th harmonics) [12], 213 nm (5th harmonics) [21]. By use of nonlinear processes such as OPO, the 1-µm NIR laser is converted to longer wavelengths [19,21,28], which extends the available laser wavelengths for SFG. Using the Er-doped fiber laser at 1550 nm together with the 1-µm NIR laser, the final target of the 193 nm laser is achieved without involving the complex and unstable nonlinear OPO process [12,31,42].

Recently, 310 mW DUV laser at 193 nm was generated by stages of SFG in the CLBO crystals. The schematic setup for this 193 nm laser was shown in Figure 2. The 193-nm laser was generated by two stages of SFG process between a 258 nm laser generated by FHG of a 1030 nm laser and a 1553 nm laser. The 1030 nm laser was based on an Yb-doped hybrid amplifier and the 1553 nm laser was from an Er-doped fiber amplifier. The 1030 nm laser and the 1553 nm laser were synchronized by a commercial available delay/digital pulse generator. The LBO crystal was utilized for the SHG to 515 nm and the CLBO crystals were used for FHG to 258 nm, SFG to 221 nm, and SFG to 193 nm. The generated 193 nm laser was operated at 6 kHz with pulse duration of 4 ns. The average power was 310 mW and the linewidth was about 4 GHz, which satisfied the needs for injection-seeding the hybrid ArF excimer laser [2,12]. As a higher-power 193 nm laser was needed to increase the signal-to-ASE-ratio of the hybrid ArF excimer laser output, a watt-level 193-nm laser has been generated due to this requirement [20].

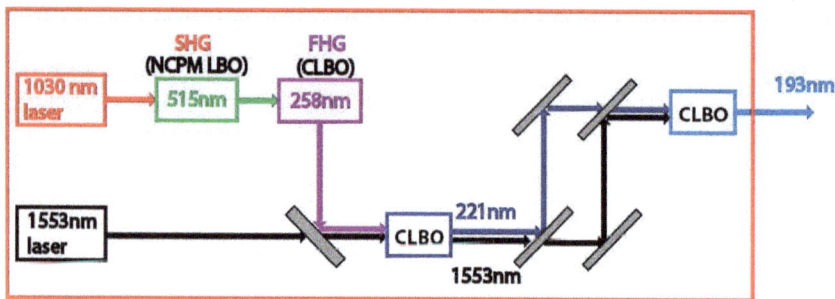

**Figure 2.** Schematic setup of 193 nm generation by frequency mixing of 1030-nm and 1553-nm lasers (NCPM: noncritical phase-matching).

## 4. DUV Laser Generation Based on Diamond Raman Laser

In this section, a new method to generate the 193 nm DUV laser is introduced, which is based on the diamond Raman laser scheme. In terms of the above methods of the 193 nm generation with 258 nm and 1553 nm, there is a limitation for its power scaling. The power of 258 nm laser is improved to more than 10 W while the power of the 1553 nm laser is limited because of the lack of the amplification medium and methods. The output power of the current Er-doped fiber amplifier is less than 2 W which is limited due to the stimulated Brillouin scattering [42]. For the solid-state laser, the average output power was also no more than 1 W [43]. Hence, this longer wavelength is usually generated by use of OPO process pumped by the Nd- or Yb-doped lasers.

In the past few decades, many Raman materials have been studied [44–46], which showed the high conversion efficiency to the first Stokes light from 1 μm. The stimulated Raman scattering (SRS) laser is remarkable for the momentum conservation automatically determined by the scattered phonons, which will induce the Stokes light beam to be a high quality [22,47]. Particularly, by use of the chemical vapor deposition (CVD) diamond, the diamond Raman laser depicted a high power result of 100 W [22] and a high slope efficiency of 84% approaching the quantum limit [23]. In detail, the most interesting advantages of diamond includes the high Raman gain (pumping at 1030 nm, $g_R = 13.5 \pm 2.0$ cm/GW) and the large frequency shift of the 1st order Raman mode (1332.5 cm$^{-1}$) [22,48], which could achieve a high-power operation and a sufficient frequency shift. Moreover, the high thermal conductivity (2200 W/m·K), which is two to three orders of magnitude higher than conventional Raman materials, and the low thermal expansion coefficient of diamond ($1.1 \times 10^{-6}$ K$^{-1}$) will be helpful for power scaling to a high level as well as its high damage threshold [45–48]. Another important advantage of the diamond is the broad transmission band. Recently, the longest solid-state Raman laser wavelength was obtained in the 3-μm region in a CVD diamond, showing a new pathway to the mid-IR laser and the possibility to cover from NIR to mid-IR wavelength [49]. Hence, by choosing a suitable Raman wavelength, a high-power 193 nm laser will be effectively generated by stages of SFG.

Figure 3 demonstrates the conceptual setup of the high-power 193 nm laser by the diamond Raman laser. The idea is also basically based on two stages of SFG process. The first DUV wavelength is at 260 nm, which is the FHG of a 1040 nm laser. Meanwhile, the wavelength of the NIR laser is 1512 nm, which is the second Stokes wavelength of a 1078 nm laser and generated by a diamond Raman laser. Both the 1040 nm and 1078 nm laser come from one dual-wavelength Yb-doped fiber laser. Each laser is amplified by stages of fiber amplifier and Yb-doped materials amplifier, respectively.

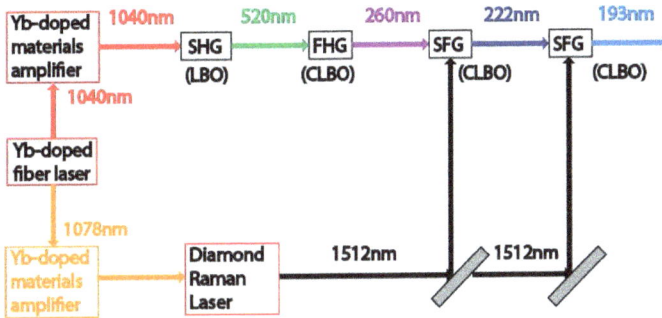

**Figure 3.** Conceptual setup of the 193 nm laser generation based on the diamond Raman laser.

The Yb-doped materials amplifier could be a $Yb:Lu_2O_3$ [50] or a $Yb^{3+}:CaCdAlO_4$ [51] laser at both wavelengths (1040 nm & 1078 nm). For the frequency conversion by nonlinear crystals to short wavelengths, the 1040 nm laser generates the green laser at 520 nm by SHG using the LBO crystal, and then the DUV laser at 260 nm will be generated by SHG from 520 nm by use of the CLBO crystal. On the other hand, the 1078 nm is the pump laser of the diamond Raman laser; its 2nd Stokes wavelength of 1512 nm will be the NIR laser for the following SFG processes to 193 nm. Hence, a DUV laser at 222 nm will be achieved by SFG between the 260-nm laser and the 1512-nm laser. The 193-nm laser will be generated by the second stage SFG between the 222-nm laser and the residual of the 1512-nm laser.

We first used a Yb:YAG laser at the wavelength of 1030 nm as the pump of the diamond Raman laser, which is the proof-of-principle experiment. For this proof-of-principle experiment, the Raman laser setup is depicted in Figure 4, of which is a short Raman laser cavity with a small piece of diamond inside it pumped by a 1030 nm laser [52]. The 1030 nm laser is from a 10 kHz Yb:YAG CTR amplifier with the average power of 27 W. The diamond is made by CVD process with a dimension of $6 \times 6 \times 1.5$ mm$^3$ and with anti-reflection (AR) coating of the pump, the 1st and the 2nd Stokes wavelengths. The beam radius of the 1030 nm pump laser is about 700 μm [52]. The diamond crystal is pumped in the direction of <110>. The Raman laser cavity is a short and linear cavity with the length of 25 mm including an input concave mirror (M1) with R = 75 mm and an output coupler (OC) with R = 50 mm. The coating of M1 is high transmission (HT) at 1030 nm and high reflection (HR) at the 1st (1194 nm) and the 2nd Stokes wavelength (1420 nm) of 1030 nm. The OC is HR for both 1030 nm and 1194 nm. The transmission of OC at 1420 nm was tried with three different values of 5%, 20%, and 50%. Two reflection mirrors (BB1-E04, Thorlabs Inc., Newton, NJ, USA) were applied to filter out the fundamental light, and the 2nd Stokes wavelength of 1420 nm was detected.

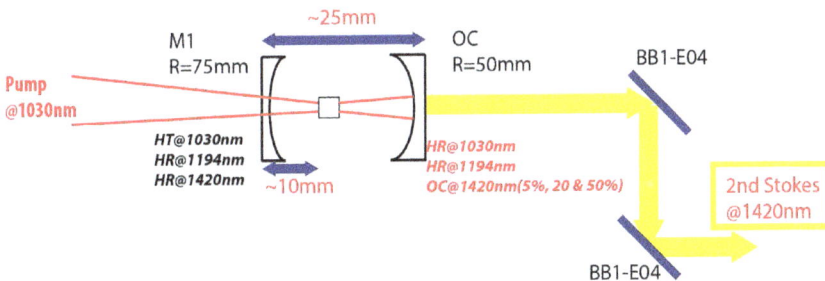

**Figure 4.** Experimental setup of the diamond Raman laser pumped by the Yb:YAG laser (HT: high transmission; HR: high reflection; OC: outup coupler).

The Raman laser achieved lasing at the wavelength of 1420 nm when the pump laser power was above the threshold power. As shown in Figure 5, the threshold power was 1.25 W and 1.5 W for the 5% and 50% OC, respectively. The threshold power for 20% OC was between 1.25 W and 1.5 W. During the cavity alignment to achieve the 2nd Stokes wavelength, the AR coating of the front surface of the diamond was easily damaged by the reflected pump beam back from the OC when the pump laser (1030 nm) power was higher than 3 W. This was because of the imperfect design of the current Raman laser cavity. Not only the AR coating of the diamond surface but also the coating of the input mirror M1 was damaged by the returned pump laser. The values of the intensity of the returned pump laser at M1 and the front surface of the diamond were estimated to be 0.6–1 J/cm$^2$. The pump power was limited to 3 W to prevent the damages inside cavity for 5% and 20% OC. For the 50% OC, the pump power was carefully increased to about 4 W during the experiment, which was still at a lower power level in order to avoid the damage of the M1 and diamond. The HT coating of the pump beam at OC or the change of the radius of the curvature of could solve this issue to realize the higher power operation. At the pump power of 3 W, the output power of the Raman laser was 130 mW, 293 mW and 325 mW using 5%, 20%, and 50% OC, respectively. By use of the 50% OC, the highest average power of this Raman laser was 0.586 W when the pump laser was 4.2 W as shown in Figure 5, with the slope efficiency of 23% which was comparable to the previous reported results. The spectrum of the generated 2nd Stokes after the two reflection mirrors was measured by a spectrometer (NIRQuest512, Ocean Optics Inc., Largo, FL, USA) as shown in Figure 6.

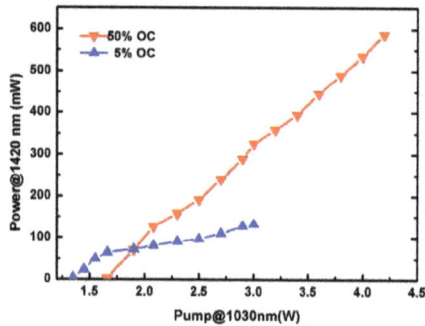

**Figure 5.** Output power of the 2nd Stokes (1420nm) vs. Pump power with 5% and 50% OC.

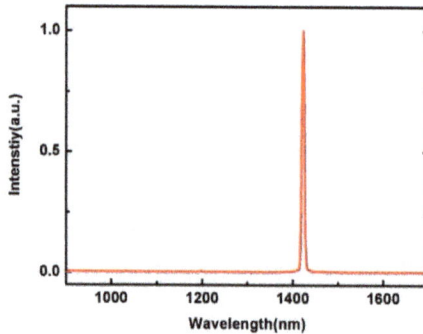

**Figure 6.** Spectrum of the diamond Raman laser at the 2nd Stokes wavelength of 1420 nm.

The pulse duration of the second Stokes light was shortened to 5 ns as shown in Figure 7a, a single-shot picture from the oscilloscope, which is a common phenomenon in Raman laser [44] while the pulse duration of the pump laser was 10 ns. The beam profile shown in Figure 7b demonstrated a good beam quality as one of the merits of Raman laser, which would be beneficial to generate a 193-nm DUV laser with a high quality beam.

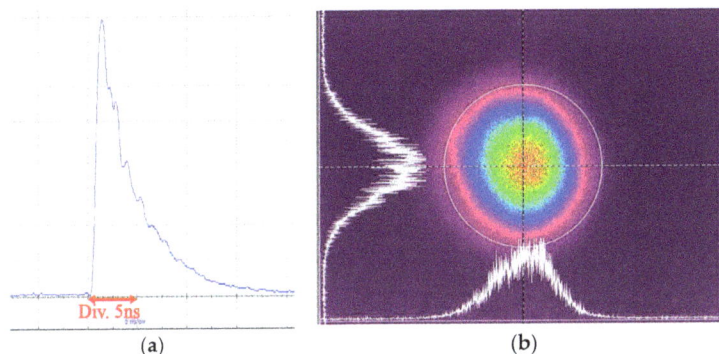

**Figure 7.** (a) Pulse duration of ~5 ns (5% OC); (b) Beam profile of the 1420 nm laser.

According to the proof-of-principle experiment, the conversion efficiency of 23% implied the possibility of achieving high-power at the second Stokes wavelength of the Raman laser. If the pump laser of 1078 nm could be obtained at the average power of 50 W, the output power of its 2nd Stokes wavelength at 1512 nm could be more than 10 W. The design of the Raman laser cavity should be optimized in order to avoid the damage of the M1 and diamond surface. For instance, the current concave OC could be replaced by a plane mirror which could mitigate the power density on the surface of M1 and the diamond. Moreover, using high damage threshold coatings for M1 and diamond would also improve the output power without any damages. Then, the NIR laser at the 2nd Stokes wavelength could be high enough, as well as its good beam quality, to generate the final DUV laser at 193 nm with a high average power.

## 5. Conclusions

We summarized the current development status of the high-power DUV laser both at the wavelength of FHG of the 1-μm laser and at 193 nm. The recent development of the solid-state/fiber laser, as well as the nonlinear optical crystals, prompts not only the average power but also the beam quality (coherence) of the DUV lasers to an unprecedented level. Moreover, combining the solid-state laser with other nonlinear process such as stimulated Raman scattering, NIR lasers could be achieved at new wavelengths with high power and high beam quality, which was tested by the proof-of-principle experiment. This Raman laser scheme provides a new route to the high-power DUV laser, especially at the wavelength of 193 nm.

**Acknowledgments:** New Energy and Industrial Technology Development Organization (NEDO).

**Conflicts of Interest:** The authors declare no conflicts of interest.

## References

1.  Allaria, E.; Castronovo, D.; Cinquegrana, P.; Craievich, P.; Dal Forno, M.; Danailov, M.B.; D'Auria, G.; Demidovich, A.; De Ninno, G.; Di Mitri, S.; et al. Two-stage seeded soft-X-ray free-electron laser. *Nat. Photonics* **2013**, *7*, 913–918. [CrossRef]

2.  Tanaka, S.; Arakawa, M.; Fuchimukai, A.; Sasaki, Y.; Onose, T.; Kamba, Y.; Igarashi, H.; Qu, C.; Tamiya, M.; Oizumi, H.; et al. Development of high coherence high power 193 nm laser. In Proceedings of the SPIE Photonics West, Solid State Lasers XXV: Technology and Devices, San Francisco, CA, USA, 13–18 February 2016; Volume 9726, p. 972625. [CrossRef]
3.  Liu, Q.; Yan, X.P.; Fu, X.; Gong, M.; Wang, D.S. High power all-solid-state fourth harmonic generation of 266 nm at the pulse repetition rate of 100 kHz. *Laser Phys. Lett.* **2009**, *6*, 203–206. [CrossRef]
4.  Délen, X.; Deyra, L.; Benoit, A.; Hanna, M.; Balembois, F.; Cocquelin, B.; Sangla, D.; Salin, F.; Didierjean, J.; Georges, P. Hybrid master oscillator power amplifier high-power narrow-linewidth nanosecond laser source at 257 nm. *Opt. Lett.* **2013**, *38*, 995–997. [CrossRef] [PubMed]
5.  Chang, C.; Krogen, P.; Liang, H.; Stein, G.J.; Moses, J.; Lai, C.; Siqueira, J.P.; Zapata, L.E.; Kärtner, F.X.; Hong, K.-H. Multi-mJ, kHz, ps deep-ultraviolet source. *Opt. Lett.* **2015**, *40*, 665–668. [CrossRef] [PubMed]
6.  Goldberg, L.; Cole, B.; McIntosh, C.; King, V.; Hays, A.D.; Chinn, S.R. Narrow-band 1 W source at 257 nm using frequency quadrupled passively Q-switched Yb:YAG laser. *Opt. Express* **2016**, *24*, 17397–17405. [CrossRef] [PubMed]
7.  Müller, M.; Klenke, A.; Gottschall, T.; Klas, R.; Rothhardt, C.; Demmler, S.; Rothhardt, J.; Limpert, J.; Tünnermann, A. High-average-power femtosecond laser at 258 nm. *Opt. Lett.* **2017**, *42*, 2826–2829. [CrossRef] [PubMed]
8.  Nishioka, M.; Fukumoto, S.; Kawamura, F.; Yoshimura, M.; Mori, Y.; Sasaki, T. Improvement of laser-induced damage tolerance in CsLiB$_6$O$_{10}$ for high-power UV laser source. In Proceedings of the Lasers and Electro-Optics/Quantum Electronics and Laser Science Conference, Baltimore, MD, USA, 1–6 June 2003. [CrossRef]
9.  Kojima, T.; Konno, S.; Fujikawa, S.; Yasui, K.; Yoshizawa, K.; Mori, Y.; Sasaki, T.; Tanaka, M.; Okada, Y. 20-W ultraviolet-beam generation by fourth-harmonic generation of an all-solid-state laser. *Opt. Lett.* **2000**, *25*, 58–60. [CrossRef] [PubMed]
10. Wang, G.; Geng, A.; Bo, Y.; Li, H.; Sun, Z.; Bi, Y.; Cui, D.; Xu, Z.; Yuan, X.; Wang, X.; et al. 28.4 W 266 nm ultraviolet-beam generation by fourth-harmonic generation of an all-solid-state laser. *Opt. Commun.* **2006**, *259*, 820–822. [CrossRef]
11. Novák, O.; Turčíčová, H.; Smrž, M.; Miura, T.; Endo, A.; Mocek, T. Picosecond green and deep ultraviolet pulses generated by a high-power 100 kHz thin-disk laser. *Opt. Lett.* **2016**, *41*, 5210–5213. [CrossRef] [PubMed]
12. Xuan, H.; Zhao, Z.; Igarashi, H.; Ito, S.; Kakizaki, K.; Kobayashi, Y. 300-mW narrow-linewidth deep-ultraviolet light generation at 193 nm by frequency mixing between Yb-hybrid and Er-fiber lasers. *Opt. Express* **2015**, *23*, 10564–10572. [CrossRef] [PubMed]
13. Xuan, H.; Qu, C.; Ito, S.; Kobayashi, Y. High-power and high-conversion efficiency deep ultraviolet (DUV) laser at 258 nm generation in the CsLiB$_6$O$_{10}$ (CLBO) crystal with a beam quality of M$^2$ < 1.5. *Opt. Lett.* **2017**, *42*, 3133–3136. [CrossRef] [PubMed]
14. Wang, L.; Zhai, N.; Liu, L.; Wang, X.; Wang, G.; Zhu, Y.; Chen, C. High-average-power 266 nm generation with a KBe$_2$BO$_3$F$_2$ prism-coupled device. *Opt. Express* **2014**, *22*, 27086–27093. [CrossRef] [PubMed]
15. Liu, Q.; Yan, X.; Gong, M.; Liu, H.; Zhang, G.; Ye, N. High-power 266 nm ultraviolet generation in yttrium aluminum borate. *Opt. Lett.* **2011**, *36*, 2653–2655. [CrossRef] [PubMed]
16. Nikitin, D.G.; Byalkovskiy, O.A.; Vershinin, O.I.; Puyu, P.V.; Tyrtyshnyy, V.A. Sum frequency generation of UV laser radiation at 266 nm in LBO crystal. *Opt. Lett.* **2016**, *41*, 1660–1663. [CrossRef] [PubMed]
17. Fang, Z.; Hou, Z.-Y.; Yang, F.; Liu, L.-J.; Wang, X.-Y.; Xu, Z.-Y.; Chen, C.-T. High-efficiency UV generation at 266 nm in a new nonlinear optical crystal NaSr$_3$Be$_3$B$_3$O$_9$F$_4$. *Opt. Express* **2017**, *25*, 26500–26507. [CrossRef] [PubMed]
18. Sakuma, J.; Kaneda, Y.; Oka, N.; Ishida, T.; Moriizumi, K.; Kusunose, H.; Furukawa, Y. Continuous-wave 193.4 nm laser with 120 mW output power. *Opt. Lett.* **2015**, *40*, 5590–5593. [CrossRef] [PubMed]
19. Jacob, J.J.; Merriam, A.J. Development of a 5-kHz solid state 193-nm actinic light source for photomask metrology and review. In Proceedings of the SPIE, 24th Annual BACUS Symposium on Photomask Technology, Monterey, CA, USA, 13–17 September 2004; Volume 5567. [CrossRef]
20. Xuan, H.; Qu, C.; Zhao, Z.; Ito, S.; Kobayashi, Y. 1 W solid-state 193 nm coherent light by sum-frequency generation. *Opt. Express* **2017**, *25*, 29172–29179. [CrossRef]

21. Mead, R.D.; Hamilton, C.E.; Lowenthal, D.D. Solid state lasers for 193-nm photolithography. In Proceedings of the SPIE, Optical Microlithography X, Santa Clara, CA, USA, 10–14 March 1997; Volume 3051. [CrossRef]
22. Williams, R.; Kitzler, O.; McKay, A.; Mildren, R. Investigating diamond Raman lasers at the 100 W level using quasi-continuous-wave pumping. *Opt. Lett.* **2014**, *39*, 4152–4155. [CrossRef] [PubMed]
23. Sabella, A.; Piper, J.A.; Mildren, R.P. 1240 nm diamond Raman laser operating near the quantum limit. *Opt. Lett.* **2010**, *35*, 3874–3876. [CrossRef] [PubMed]
24. Nubbemeyer, T.; Kaumanns, M.; Ueffing, M.; Gorjan, M.; Alismail, A.; Fattahi, H.; Brons, J.; Pronin, O.; Barros, H.G.; Major, Z.; et al. 1 kW, 200 mJ picosecond thin-disk laser system. *Opt. Lett.* **2017**, *42*, 1381–1384. [CrossRef] [PubMed]
25. Rothhardt, J.; Rothhardt, C.; Müller, M.; Klenke, A.; Kienel, M.; Demmler, S.; Elsmann, T.; Rothhardt, M.; Limpert, J.; Tünnermann, A. 100 W average power femtosecond laser at 343 nm. *Opt. Lett.* **2016**, *41*, 1885–1888. [CrossRef] [PubMed]
26. Lesparre, F.; Martial, I.; Didierjean, J.; Gomes, J.T.; Pallmann, W.; Resan, B.; Loescher, A.; Negel, J.-P.; Graf, T.; Ahmed, M.A.; et al. High-power Yb:YAG single-crystal fiber amplifiers for femtosecond lasers. In Proceedings of the SPIE Photonics West, Solid State Lasers XXV: Technology and Devices, San Francisco, CA, USA, 7–12 February 2015; Volume 9432. [CrossRef]
27. Scholz, M.; Opalevs, D.; Leisching, P.; Kaenders, W.; Wang, G.; Wang, X.; Li, R.; Chen, C. A bright continuous-wave laser source at 193 nm. *Appl. Phys. Lett.* **2013**, *103*, 051114. [CrossRef]
28. Zhang, H.; Wang, G.; Guo, L.; Geng, A.; Bo, Y.; Cui, D.; Xu, Z.; Li, R.; Zhu, Y.; Wang, X.; et al. 175 to 210 nm widely tunable deep-ultraviolet light generation based on KBBF crystal. *Appl. Phys. B* **2008**, *93*, 323–326. [CrossRef]
29. Kanai, T.; Wang, X.; Adachi, S.; Watanabe, S.; Chen, C. Watt-level tunable deep ultraviolet light source by a KBBF prism-coupled device. *Opt. Express* **2009**, *17*, 8696–8703. [CrossRef] [PubMed]
30. Ito, S.; Onose, T.; Watanabe, S.; Kanai, T.; Kakizaki, K.; Matsunaga, T.; Chen, C.; Kobayashi, Y.; Zhou, C.; Fujimoto, J.; et al. A Sub-Watt, Line-narrowing, 193-nm Solid State Laser Operating at 6 kHz with KBBF for Injection-locking ArF excimer laser systems. In Proceedings of the Lasers, Sources, and Related Photonic Devices, San Diego, CA, USA, 29 January–3 February 2012. [CrossRef]
31. Kawai, H.; Tokuhisa, A.; Doi, M.; Miwa, S.; Matsuura, H.; Kitano, H.; Owa, S. UV light source using fiber amplifier and nonlinear wavelength conversion. In Proceedings of the Lasers and Electro-Optics/Quantum Electronics and Laser Science Conference, Baltimore, MD, USA, 1–6 June 2003.
32. Nakazato, T.; Tsuboi, M.; Onose, T.; Tanaka, Y.; Sarukura, N.; Ito, S.; Kakizaki, K.; Watanabe, S. Development of high coherence, 200 mW, 193 nm solid-state laser at 6 kHz. In Proceedings of the SPIE Photonics West, Solid State Lasers XXV: Technology and Devices, San Francisco, CA, USA, 7–12 Februray 2015; Volume 9342. [CrossRef]
33. Umemura, N.; Ando, M.; Suzuki, K.; Takaoka, E.; Kato, K.; Hu, Z.; Yoshimura, M.; Mori, Y.; Sasaki, T. 200-mW-average power ultraviolet generation at 0.193 μm in $K_2Al_2B_2O_7$. *Appl. Opt.* **2003**, *42*, 2716–2719. [CrossRef] [PubMed]
34. Ringling, J.; Kittelmann, O.; Seifert, F.; Noack, F.; Korn, G.; Squier, J.A. Femtosecond solid state light sources tunable around 193 nm. In Proceedings of the SPIE, Generation, Amplification, and Measurement of Ultrashort Laser Pulses, Los Angeles, CA, USA, 23–29 January 1994; Volume 2116. [CrossRef]
35. Tran, T.T.; Yu, H.; Rondinelli, J.M.; Poeppelmeier, K.R.; Halasyamani, P.S. Deep Ultraviolet Nonlinear Optical Materials. *Chem. Mater.* **2016**, *28*, 5238–5258. [CrossRef]
36. Yang, Y.; Jiang, X.; Lin, Z.; Wu, Y. Borate-Based Ultraviolet and Deep-Ultraviolet Nonlinear Optical Crystals. *Crystals* **2017**, *7*, 95. [CrossRef]
37. Trabs, P.; Noack, F.; Aleksandrovsky, A.S.; Zaitsev, A.I.; Petrov, V. Generation of coherent radiation in the vacuum ultraviolet using randomly quasi-phase-matched strontium tetraborate. *Opt. Lett.* **2016**, *41*, 618–621. [CrossRef] [PubMed]
38. Nakazato, T.; Ito, I.; Kobayashi, Y.; Wang, X.; Chen, C.; Watanabe, S. 149.8 nm, the shortest wavelength generated by phase matching in nonlinear crystals. In Proceedings of the SPIE Photonics West, Nonlinear Frequency Generation and Conversion: Materials and Devices XVI, San Francisco, CA, USA, 28 January–2 February 2017; Volume 10088, p. 1008804. [CrossRef]
39. Nakazato, T.; Ito, I.; Kobayashi, Y.; Wang, X.; Chen, C.; Watanabe, S. Phase-matched frequency conversion below 150 nm in $KBe_2BO_3F_2$. *Opt. Express* **2016**, *24*, 17149–17158. [CrossRef] [PubMed]

40. Liu, G.; Wang, G.; Zhu, Y.; Zhang, H.; Zhang, G.; Wang, X.; Zhou, Y.; Zhang, W.; Liu, H.; Zhao, L.; et al. Development of a vacuum ultraviolet laser-based angle-resolved photoemission system with a super high energy resolution better than 1 meV. *Rev. Sci. Instrum.* **2008**, *79*, 023105. [CrossRef] [PubMed]

41. Chen, C.; Sasaki, T.; Li, R.; Wu, Y.; Lin, Z.; Mori, Y.; Hu, Z.; Wang, J.; Uda, S.; Yoshimura, M.; et al. *Nonlinear Optical Borate Crystals: Principles and Applications*; Wiley-VCH Verlag & Co. KGaA: Weinheim, Germany, 2012; pp. 164–166. ISBN 978-3-527-41009-5.

42. Zhao, Z.; Xuan, H.; Igarashi, H.; Ito, S.; Kakizaki, K.; Kobayashi, Y. Single frequency, 5 ns, 200 µJ, 1553 nm fiber laser using silica based Er-doped fiber. *Opt. Express* **2015**, *23*, 29764–29771. [CrossRef] [PubMed]

43. Tolstik, N.A.; Kurilchik, S.V.; Kisel, V.E.; Kuleshov, N.V.; Maltsev, V.V.; Pilipenko, O.V.; Koporulina, E.V.; Leonyuk, N.I. Efficient 1 W continuous-wave diode-pumped Er,Yb:YAl3(BO3)4 laser. *Opt. Lett.* **2007**, *32*, 3233–3235. [CrossRef] [PubMed]

44. Pask, H.M. The design and operation of solid-state Raman lasers. *Prog. Quantum Electron.* **2003**, *27*, 3–56. [CrossRef]

45. Zverev, P.G.; Basiev, T.T.; Osiko, V.V.; Kulkov, A.M.; Voitsekhovskii, V.N.; Yakobson, V.E. Physical, chemical and optical properties of barium nitrate Raman crystal. *Opt. Mater.* **1999**, *11*, 315–334. [CrossRef]

46. Kaminskii, A.A.; McCray, C.L.; Lee, H.R.; Lee, S.W.; Temple, D.A.; Chyba, T.H.; Marsh, W.D.; Barnes, J.C.; Annanenkov, A.N.; Legun, V.D.; et al. High efficiency nanosecond Raman lasers based on tetragonal PbWO$_4$ crystals. *Opt. Commun.* **2000**, *183*, 277–287. [CrossRef]

47. McKay, A.; Kitzler, O.; Mildren, R.P. Simultaneous brightness enhancement and wavelength conversion to the eye-safe region in a high-power diamond Raman laser. *Laser Photonics Rev.* **2014**, *8*, L37–L41. [CrossRef]

48. Feve, J.-P.M.; Shortoff, K.E.; Bohn, M.J.; Brasseur, J.K. High average power diamond Raman laser. *Opt. Express* **2011**, *19*, 913–922. [CrossRef] [PubMed]

49. Sabella, A.; Piper, J.A.; Mildren, R.P. Diamond Raman laser with continuously tunable output from 3.38 to 3.80 µm. *Opt. Lett.* **2014**, *39*, 4037–4040. [CrossRef] [PubMed]

50. Weichelt, B.; Wentsch, K.S.; Voss, A.; Ahmed, M.A.; Graf, T. A 670 W Yb:Lu$_2$O$_3$ thin-disk laser. *Laser Phys. Lett.* **2012**, *9*, 110–115. [CrossRef]

51. Boudeile, J.; Druon, F.; Hanna, M.; Georges, P.; Zaouter, Y.; Cormier, E.; Petit, J.; Goldner, P.; Viana, B. Continuous-wave and femtosecond laser operation of Yb:CaGdAlO$_4$ under high-power diode pumping. *Opt. Lett.* **2007**, *32*, 1962–1964. [CrossRef] [PubMed]

52. Xuan, H.; Ito, S.; Kobayashi, Y. High power Yb:YAG ceramics laser and diamond Raman laser for frequency conversion to DUV. In Proceedings of the 8th International Symposium on Ultrafast Phenomena and Terahertz Waves, Chongqing, China, 10–12 October 2016. [CrossRef]

*applied
sciences*

MDPI

*Article*

# A High-Energy, 100 Hz, Picosecond Laser for OPCPA Pumping

Hongpeng Su [1,2], Yujie Peng [1,*], Junchi Chen [1], Yanyan Li [1], Pengfei Wang [1,2] and Yuxin Leng [1,*]

[1]   State Key Laboratory of High Field Physics, Shanghai Institute of Optics and Fine Mechanics,
     Chinese Academy of Sciences, 390# Qinghe Road, Jiading District, Shanghai 201800, China;
     spamsu@siom.ac.cn (H.S.); chenjunchi01@163.com (J.C.); yyli@siom.ac.cn (Y.L.);
     wangpengfei@siom.ac.cn (P.W.)
[2]   University of Chinese Academy of Sciences, Beijing 100190, China
*   Correspondence: yjpeng@siom.ac.cn (Y.P.); lengyuxin@mail.siom.ac.cn (Y.L.);
     Tel.: +86-021-6991-8261 (Y.P.); +86-021-6991-8436 (Y.L.)

Received: 31 August 2017; Accepted: 25 September 2017; Published: 27 September 2017

**Abstract:** A high-energy diode-pumped picosecond laser system centered at 1064 nm for optical parametric chirped pulse amplifier (OPCPA) pumping was demonstrated. The laser system was based on a master oscillator power amplifier configuration, which contained an Nd:YVO$_4$ mode-locked seed laser, an LD-pumped Nd:YAG regenerative amplifier, and two double-pass amplifiers. A reflecting volume Bragg grating with a 0.1 nm reflective bandwidth was used in the regenerative amplifier for spectrum narrowing and pulse broadening to suit the pulse duration of the optical parametric amplifier (OPA) process. Laser pulses with an energy of 316.5 mJ and a pulse duration of 50 ps were obtained at a 100 Hz repetition rate. A top-hat beam distribution and a 0.53% energy stability (RMS) were achieved in this system.

**Keywords:** diode-pumped solid-state laser; picosecond laser; OPCPA pumping

---

## 1. Introduction

High energy, ultrafast laser in the mid-infrared (mid-IR) is desirable for applications in strong field physics [1], high-harmonic generation [2], and driving X-ray sources [3]. An optical parametric chirped pulse amplifier (OPCPA) is one of the most promising ways to achieve amplification for mid-IR laser pulses by applying proper nonlinear crystals and a phase matching condition. A kHz, mJ-level OPCPA at 2.1 μm is reported with a high-energy cryogenic Yb:YAG pump laser [4]. The current limitation for a higher OPCPA output depends upon the development of high-energy picosecond pumping sources. Several investigations have been developed for the generation of high-energy OPCPA pump sources. A picosecond pump laser for IR OPCPA delivering 25 mJ at 3 kHz based on a Yb:YAG thin-disk amplifier has been reported [5]. A 100 mJ, 1 kHz laser output has been achieved as a pump for picosecond OPCPA based on Yb:YAG thin disk regenerative amplifier [6]. Additionally, an OPCPA pump laser based on rod-shaped Nd:YAG crystals has been reported, producing 130 mJ, 64 ps pulses with a repetition of 300 Hz [7].

In this paper, we present a diode-pumped solid-state laser (DPSSL) amplification system based on a master oscillator power amplifier (MOPA) configuration, delivering 316.5 mJ, 50 ps pulse energy at a wavelength of 1064 nm and a repetition of 100 Hz. The amplification system works at room temperature. A thermal lens and thermally induced depolarization were studied and compensated. A reflecting volume Bragg grating was utilized in the regenerative amplifier to achieve temporal stretching of the laser pulses to suit the width of the signal laser pulses in the optical parametric amplifier (OPA) process. The output of the amplification system also features an approximately flat-top spatial distribution and a 0.53% energy stability (RMS).

## 2. Experiments and Results

### 2.1. Front-End and Regenerative Amplifier

The schematic of the laser system is shown in Figure 1. The master oscillator is a home-built Nd:YVO$_4$ mode-locked laser, providing up to 280 mW, 8.4 ps laser pulses at a wavelength of 1064 nm and a frequency of 80 MHz. The OPCPA system we serve is seeded by a commercial ultrafast laser (Vitara, Coherent Corporation, Santa Clara, CA, United States) centered at 800 nm. The laser is then converted to 4 µm via a three-stage OPA based on KTiOAsO$_4$ (KTA) nonlinear crystals [8]. The 4 µm laser will serve as the signal pulses in the following OPCPA process. OPA is an interactive process between pump pulses and signal pulses, thus the temporal and spatial overlap between two pulses determines actual gain performance. To achieve perfect temporal synchronization, the Nd:YVO$_4$ mode-locked oscillator is designed to operate at 80 MHz, which is the same with the commercial 800 nm laser. The commercial 800 nm oscillator is equipped with a synchronization accessory, which can synchronize its oscillator to an external Radio Frequency (RF) source. Part of the Nd:YVO$_4$ mode-locked laser pulses are utilized and converted into a master signal by a photo-diode. The synchronization accessory will accept the trigger and automatically adjust the length of the cavity to achieve precise frequency synchronization with the master Nd:YVO$_4$ oscillator. At the same time, the trigger signal from the Nd:YVO$_4$ mode-locked oscillator is transmitted to a synchronizer, which provides all delay signals for latter electrical devices.

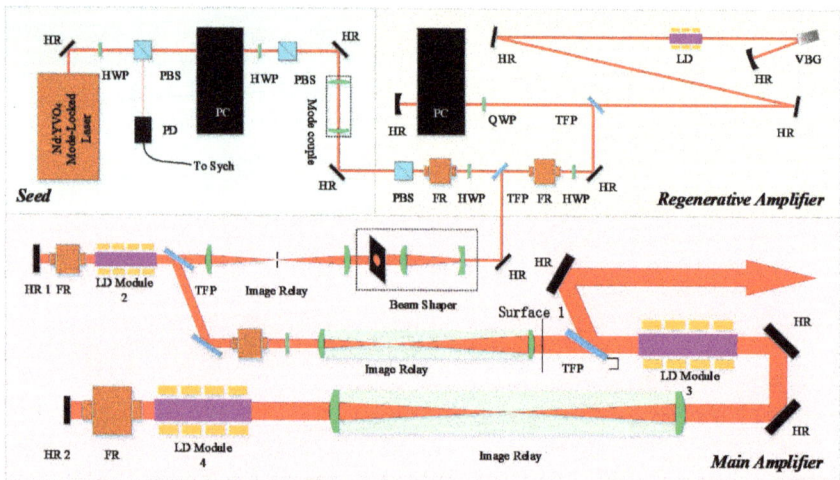

**Figure 1.** Schematic of the laser system. HR: high reflector; HWP: half-wave plate; PBS: polarization beam splitter; PD: photo diode; PC: Pockels cell; FR: Faraday rotator; TFP: thin-film plate; QWP: quarter-wave plate; LD: laser diode; VBG: volume Bragg grating.

Pulses from the Nd:YVO$_4$ mode-locked oscillator first pass through a half-wave plate (HWP), which works with a polarization beam splitter (PBS), to control the pulse energy injected into the amplification system by rotating the HWP. A Pockels cell (PC) is operated at half-wave voltage with a repetition of 100 Hz, working together with two PBSs and an HWP to achieve pulse picking and isolation from the damage of returning pulses. When the half-wave voltage is applied to the Pockels cell, pulses will maintain p-polarization after passing through the Pockels cell and the HWP. When the voltage is off, laser pulses will change their polarization after passing through the HWP and exit from either PBS. Before they are injected into the regenerative amplifier, laser pulses of 100 Hz are aligned and expanded to 1.8 mm with a pair of plano-convex lenses, which suit the cavity mode of the regenerative amplifier.

Before seeding the regenerative amplifier, laser pulses pass through an optical isolator composed of a PBS, an HWP, and a Faraday rotator (FR) to ensure extra protection for the oscillator. The regenerative amplifier has a linear cavity with a stable cavity mode. The diode-pumped amplification module (Cutting Edge Optronics, RBAT30-1P, St. Charles, MO, USA) has a rod-shaped Nd:YAG crystal with a diameter of 3 mm and a length of 60 mm. When operated at 180 A and 300 Hz with a pulse width of 250 μs, this module has a stored energy of 197.7 mJ. In this system, the operating parameters are set at 75 A and 100 Hz with a duration of 250 μs, which matches the upper state lifetime of the Nd:YAG gain material. A Pockels cell is operated at quarter-wave voltage with an operation frequency of 100 Hz, whose time delay is precisely controlled by the synchronizer to achieve correct build-up of the pulse energy in the cavity before exiting.

To achieve better temporal overlap between pump pulses and signal pulses in the OPA process, pump pulses should also be stretched to tens of picoseconds to suit the temporal width of signal pulses. In this work, a reflecting volume Bragg grating (VBG) is inserted as one of the mirrors to achieve the temporal stretching. The VBG we utilize (OptiGrate, RBG-1064-99, Oviedo, FL, USA) has an aperture of 8.0 mm × 5.0 mm and a thickness of 12 mm. The VBG has a bandwidth of 0.1 nm centered at 1064 nm with a reflection efficiency as high as 99.7%. Narrower reflected bandwidth will result in stretching in the time-domain. To obtain a maximum, saturated output with a fixed round-trip time, the input pulse energy needs to be maintained constant. If we want to change the duration of the output pulse, we need to vary the energy of the injected pulse and make the regenerative amplifier at the saturated state with different round-trips. We precisely control the input pulse energy, and after 20 round trips in the regenerative cavity, laser pulses are stretched to 115 ps, which is shown in Figure 2a. Using stretched pulses has two advantages: on the one hand, it will match the width of the signal pulses in the OPA process; on the other hand, stretched pulses lower the peak power of laser pulses and promise a higher energy gain without bringing about damage to optical devices. After 20 round trips, a total 2.2 mJ of pulse energy is achieved via the regenerative amplifier. The beam distribution is shown in Figure 2b.

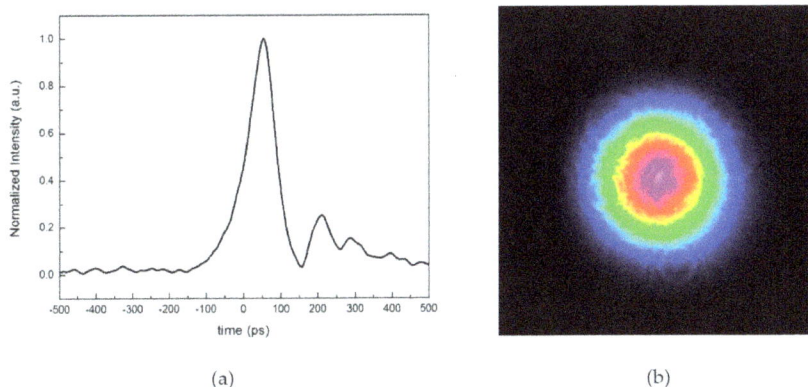

(a)                                                                    (b)

**Figure 2.** (a) Temporal profile and (b) beam distribution of amplified pulses from the regenerative amplifier.

The laser pulses of the regenerative amplifier are then expanded to 6.5 mm and aligned with a pair of plano-convex and plano-concave lenses. In the OPA process, a top-hat distribution of pump pulses is preferred due to its uniform amplification gain. To achieve a top-hat distribution of the laser beam, a serrated aperture with a diameter of 6 mm is inserted behind the regenerative amplifier to select the uniform part of the laser beam, which also reduces the pulse energy from 2.2 to 1.6 mJ.

The beam at the serrated aperture is then relay-imaged to the high-reflection mirror 1 (HR1) with a pair of plano-convex lenses. A pinhole is set at the focal point in the relay-imaging setup to achieve spatial filtering.

### 2.2. Power Amplifiers and Compensation of Thermal Effects

The power amplification stage consists of two successive laser gain configurations. The first setup contains a diode-pumped gain module (Cutting Edge Optronics, REA6306-1P), which has a rod-shaped Nd:YAG crystal with a diameter of 6 mm and a length of 120 mm. This module has a maximum operating parameter of 180 A and 300 Hz, a condition under which it has a stored energy of 513.3 mJ. In this system, the module is operated at 100 A and 100 Hz with a small signal gain of 3.49 in consideration of controllable thermal effects and sufficient amplification performance. After exiting from the serrated aperture, laser pulses pass through the gain module and a Faraday rotator, and are reflected back. The laser pulses change their polarization after the double-pass of the Faraday rotator and exit from thin film polarizer (TFP).

In the diode-pumped gain module, unabsorbed pumping light will bring about heat accumulation. A water-cooling circulatory system is used to take away the redundant heat to prevent the fracture of the gain crystal. The pumping system together with a water-cooling system will result in a nonuniform radial temperature distribution in the rod-shaped crystal. The thermal strains in the rod-shaped crystal brought by the nonuniform radial temperature distribution will result in a change in the refractive index via the photoelastic effect, which will lead to a thermally induced birefringence and thermal lensing.

The thermally induced birefringence will bring about depolarization and lead to a loss of laser energy when pulses pass through polarization-related devices. According to Reference [9], thermally induced depolarization will be compensated when laser pulses pass through the same gain module with a 90-degree polarization rotation. Thus, a Faraday rotator is placed between the gain module and HR1 to achieve the polarization rotation and depolarization compensation. With our measurements, the loss of depolarization is below 3%.

After the double-pass of Gain Module 2, laser pulses are relay-imaged from HR1 to Surface 1 and expanded from 6 to 10 mm with an imaging system composed of a pair of plano-convex lenses. Laser pulses will show little convergence due to a thermal lensing effect in Gain Module 2. When operated at 100 Hz and 100 A, the gain module has a measured focal length of 30 m, as shown in Figure 3. Thus, in the imaging setup, the distance between two lenses is precisely adjusted to achieve the compensation of the thermal lensing effect and alignment for the laser beam. In order to avoid air breakdown, a vacuum tube is placed in the imaging setup. An HWP and Faraday rotator are placed inside the imaging setup together with two TFPs to achieve another optical isolation. After the double-pass of Gain Module 2, laser pulses are amplified to 38.5 mJ.

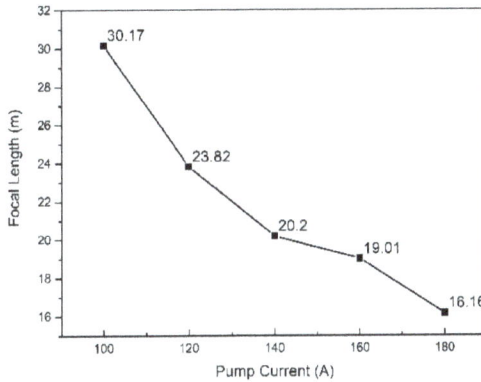

**Figure 3.** The relationship between pump current and the focal length of the gain module.

The second part of the power amplification setup consists of two identical gain modules (Cutting Edge Optronics, REA12006-3P), which has a rod-shaped Nd:YAG crystal with a diameter of 10 mm and a length of 120 mm. Each has a stored energy of 1.05 J when operated at 125 A and 300 Hz. Both gain modules are operated at 100 A and at 100 Hz with a small signal gain of 2.88 in this system. An equivalent focal length of 3.2 m is measured in such an operating situation. To obtain the best beam distribution in the exit, the laser beam should be relay-imaged from Surface 1 to HR2 and then to the exit. We take advantage of the thermal lensing of the gain module. A plano-convex lens with a focal length of 500 mm is placed behind with a distance of 400 mm to the center of the gain module. After passing through the gain module and the plano-convex lens, laser pulses converge at the focus point. Another identical gain module and a plano-convex lens are placed symmetrically with the focus point. Thermal lenses in two gain modules together with two plano-convex lenses consist of an equivalent imaging system. Due to this equivalent imaging system, laser pulses will pass through the same gain modules with the same optical path in the reflecting round. To overcome thermally induced depolarization, a Faraday rotator is placed behind Module 4 to achieve the polarization rotation and depolarization compensation.

Via the total power amplification system, laser pulses are amplified to 316.5 mJ. An approximate top-hat beam distribution with near-field modulation (peak to mean) of 1.46 is achieved, as shown in Figure 4a. A 10 min energy stability has thus been tested, and 0.53% RMS is achieved, as shown in Figure 4b.

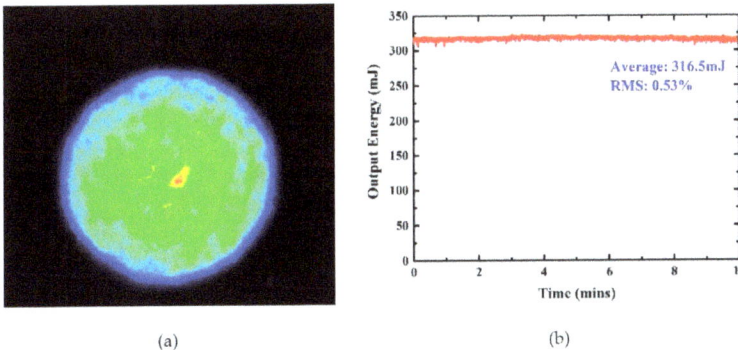

(a)

(b)

**Figure 4.** (**a**) Beam distribution and (**b**) energy stability of amplified pulses from the whole amplifier.

Due to the gain saturation effect of the Nd:YAG amplifier, after two successive double-pass gain modules, the pulse temporal width shortened from 115 to 50 ps, as shown in Figure 5, which is suitable for the signal laser pulses in the OPA process.

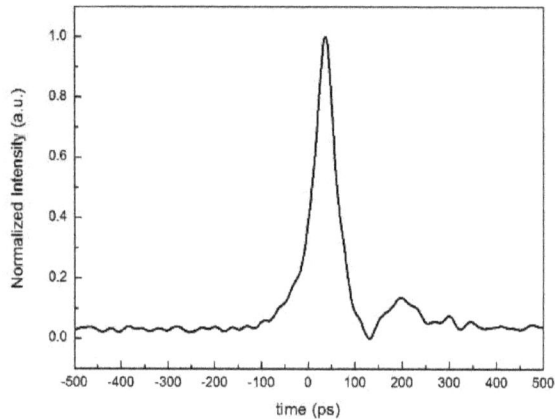

**Figure 5.** Pulse temporal width via the amplifying system.

## 3. Conclusions

In conclusion, a laser amplification system combined with an Nd:YVO$_4$ seed, a regenerative amplifier based on an Nd:YAG crystal, and two double-pass Nd:YAG power amplifiers is demonstrated. Thermally induced birefringence and thermal lensing are compensated. A 316.5 mJ, 50 ps laser pulse is achieved with an operating frequency of 100 Hz. The amplification system also features an approximately flat-top distribution and a 0.53% energy stability (RMS). This high-repetition-rate picosecond laser could be used for OPCPA pumping and other applications.

**Acknowledgments:** This work is supported by the National Science Foundation of China (NSFC) (11127901, 11134010), Shanghai Sailing Program (15YF1413500), the Strategic Priority Research Program of the Chinese Academy of Sciences, rant No. XDB1603, International S&T Cooperation Program of China, Grant No. 2016YFE0119300.

**Author Contributions:** Yanyan Li conceived and designed the experiments; Hongpeng Su and Yuxin Leng performed the experiments; Hongpeng Su and Yujie Peng analyzed the data; Hongpeng Su, Yujie Peng, Junchi Chen, Yanyan Li, Pengfei Wang and Yuxin Leng contributed reagents/materials/analysis tools; Hongpeng Su wrote the paper.

## References

1. Wolter, B.; Pullen, M.G.; Baudisch, M.; Sclafani, M.; Hemmer, M.; Senftleben, A.; Schröter, C.D.; Ullrich, J.; Moshammer, R.; Biegert, J. Strong-Field Physics with Mid-IR Fields. *Phys. Rev. X* **2015**, *5*, 021034. [CrossRef]
2. Hong, K.-H.; Lai, C.-J.; Siqueira, J.P.; Krogen, P.; Moses, J.; Chang, C.-L.; Stein, G.J.; Zapata, L.E.; Kärtner, F.X. Multi-mJ, kHz, 2.1 μm optical parametric chirped-pulse amplifier and high-flux soft X-ray highharmonic generation. *Opt. Lett.* **2014**, *39*, 3145–3148. [CrossRef] [PubMed]
3. Popmintchev, T.; Chen, M.-C.; Popmintchev, D.; Arpin, P.; Brown, S.; Ališauskas, S.; Andriukaitis, G.; Balčiunas, T.; Mücke, O.D.; Pugzlys, A.; et al. Bright coherent ultrahigh harmonics in the keV X-ray regime from mid-infrared femtosecond lasers. *Science* **2012**, *336*, 1287–1291. [CrossRef] [PubMed]
4. Hong, K.; Huang, S.; Moses, J.; Fu, X.; Lai, C.; Cirmi, G.; Sell, A.; Granados, E.; Keathley, P.; Kärtner, F. High-energy, phase-stable, ultrabroadband kHz OPCPA at 2.1 μm pumped by a picosecond cryogenic Yb:YAG laser. *Opt. Express* **2011**, *19*, 15538–15548. [CrossRef] [PubMed]

5.  Metzger, T.; Schwarz, A.; Teisset, C.Y.; Sutter, D.; Killi, A.; Kienberger, R.; Krausz, F. High-repetition-rate picosecond pump laser based on a Yb:YAG disk amplifier for optical parametric amplification. *Opt. Lett.* **2009**, *34*, 2123–2125. [CrossRef] [PubMed]

6.  Novák, J.; Bakule, P.; Green, J.T.; Hubka, Z.; Rus, B. 100 mJ thin disk regenerative amplifier at 1 kHz as a pump for picosecond OPCPA. In Proceedings of the 2015 Conference on Lasers and Electro-Optics (CLEO), San Jose, CA, USA, 10–15 May 2015.

7.  Noom, D.W.E.; Witte, S.; Eikema, K.S.E. High-energy, high-repetition-rate picosecond pulses from a quasi-CW diode-pumped Nd:YAG system. *Opt. Lett.* **2013**, *38*, 3021–3023. [CrossRef] [PubMed]

8.  Chen, Y.; Li, Y.; Li, W.; Guo, X.; Leng, Y. Generation of high beam quality, high-energy and broadband tunable mid-infrared pulse from a KTA optical parametric amplifier. *Opt. Commun.* **2016**, *365*, 7–13. [CrossRef]

9.  Lü, Q.; Kugler, N.; Weber, H.; Dong, S.; Müller, N.; Wittrock, U. A novel approach for compensation of birefringence in cylindrical Nd:YAG rods. *Opt. Quantum Electron.* **1996**, *28*, 57–69. [CrossRef]

*applied*
*sciences*

MDPI

*Article*

# Temporally Programmable Hybrid MOPA Laser with Arbitrary Pulse Shape and Frequency Doubling

Mingming Nie , Qiang Liu *, Encai Ji , Xuezhe Cao and Xing Fu

State Key Laboratory of Precision Measurement Technology and Instruments, Department of Precision Instrument, Tsinghua University, Beijing 100084, China; nmm100100@163.com (M.N.); jiencaihit@163.com (E.J.); xuezhecao@gmail.com (X.C.); fuxing@mail.tsinghua.edu.cn (X.F.)
* Correspondence: qiangliu@mail.tsinghua.edu.cn; Tel.: +86-130-2104-5699

Received: 6 August 2017; Accepted: 28 August 2017; Published: 1 September 2017

Featured Application: **The proposed device can be very useful to precision manufacturing, Lidar and freespace communication.**

Abstract: An arbitrary pulse shape by compensating gain saturation in a solid-state Master oscillator power amplifier (MOPA) system made up of three Neodymium doped yttrium vanadate (Nd:YVO$_4$) amplifiers is demonstrated. By investigating the amplifier dynamics in detail, car-shaped pulse shapes were obtained with compensated pulse distortion. Desired pulse shapes, such as multiple-step, square, parabolic, and Gaussian pulses, were achieved, with a high peak power level of 41.6 kW and a narrow linewidth less than 0.06 nm. In addition, through second harmonic generation (SHG), a green laser with different pulse shapes was obtained, with a maximum conversion efficiency of 42.6%.

Keywords: laser amplifiers; diode-pumped; pulsed laser; pulse shaping

## 1. Introduction

In the fields of optical sensing and Lidar applications, a long pulse duration, a good pulse shape, as well as a narrow spectral linewidth are required to achieve high precision measurements [1–5]. Master oscillator power amplifier (MOPA) systems seeded by a modulated semiconductor diode is a cost-effective approach to obtaining the above parameters. However, the gain saturation effect of amplifiers would cause pulse shape distortion and would shorten the pulse duration in the case of deep saturation. To obtain the required laser parameters for the above-mentioned applications, the negative influence of pulse distortion from MOPA systems should be eliminated. Therefore, the inverse problem on how to obtain the desirable output waveform by actively controlling the input pulse shape and compensating the gain saturation effect should be solved.

The general compensating method was firstly proposed by Siegman in 1986 in his book Lasers [6]. The key of the method is to get the function $G(t)$ describing the change of the input waveform, which is directly related to the gain characteristics of the amplifier. Many researchers have reported works on the pre-compensation of pulse distortion in master oscillator fiber power amplifier (MOFPA) systems [7–11]. Although it is easy to obtain $G(t)$ and control the pulse shape in fiber amplifiers, the peak power is limited by the nonlinear effect in the fiber [5,12], especially for the case of narrow spectral linewidth.

To overcome the shortage of fiber amplifiers, solid-state amplifiers are introduced in this paper. From the literature investigation, for solid-sate amplifiers, few have proposed a way of obtaining $G(t)$ and arbitrary output waveforms accurately in theory. Few works have demonstrated compensation of pulse distortion in solid-state amplifiers by experiment. With the help of a complicated computational system Laser Performance Operations Model (LPOM), a square pulse for a huge laser system was achieved by the group of National Ignition Facility (NIF) [13]. However, the detailed information of the software is not provided by the group. Other software packages for the simulation of the output from an amplification chain include Miró [14] and GLAD [15]. However, they are not easy to access for a high price.

In 2015, Lu Xu et al. [16] demonstrated an experimental method of shaping the input pulse of a solid-state MOPA system by introducing a device made up of two polarizers, one Pockels cell and one half-wave plate. However, the shaping ability is restricted to providing arbitrary input pulse, and the device could not compensate the pulse distortion in the case of an arbitrary saturation degree. Although a square pulse shape is demonstrated in the experiment, the difference between the desired pulse shape and the measured one is large due to the non-linear edge of the input pulse. Gain characteristics and pulse shape transfer function are not deduced theoretically for their MOPA system in that paper either. Table 1 shows the detailed information of progress in temporal compensation for saturated amplifiers.

Table 1. Progress in temporal compensation for saturated amplifiers.

| Amplifier | Peak Power (kW) | Linewidth (nm) | Exp or Theo [1] | Arbitrary Waveform | Reference |
|---|---|---|---|---|---|
| | 2.6 | 3 | theo | yes | [7] |
| fiber | 1.5 | - | theo | yes | [8] |
| | 18 | 0.3 | exp | yes | [9] |
| solid-state | - | - | exp | yes | [13] |
| | $1.34 \times 10^8$ | narrow | exp | no | [16] |

[1] The method of obtaining the arbitrary waveforms: experimental or theoretical.

Our previous work [17] emphasizes the demonstration of an arbitrary burst envelope for an MOPA laser operating in burst mode, which could be used in chirped amplification. In this study, we obtained arbitrary nanosecond pulse shapes in a solid-state MOPA system made up of three Neodymium doped yttrium vanadate (Nd:YVO$_4$) amplifiers. The theoretical method is similar to that in [17]. The pulse distortion was experimentally compensated without a feedback loop. Several particular desired pulse shapes were demonstrated, with a high peak power level and a narrow spectral linewidth. The MOPA laser, compared with the fiber laser, achieved a higher pulse energy and peak power by introducing more solid-state amplifiers, without the limit of a nonlinear effect. Moreover, second harmonic generation (SHG) was demonstrated after the MOPA source, resulting in arbitrary waveforms of the green laser. Based on the experimental results, how to generate desired pulse shapes through the SHG process is discussed.

## 2. MOPA Setup

A schematic diagram of the experiment setup is depicted in Figure 1. Figure 1a shows the setup of the fiber stage, including a distributed feedback (DFB) laser with fiber-coupled output and two fiber amplifiers. The setup of the fiber stage was similar to that in [17]. The difference was that electro-optic modulator (EOM2) was replaced by an acousto-optic modulator (AOM, Gooch & Housego T-M200-0.1C2G-3-F2P, llminster, UK). The maximum rise/fall time of the AOM is 10 ns, and the maximum modulation frequency is 200 MHz. Its function was enhancing the extinction ratio and pulse contrast of the output pulses.

The photo of the setup of the solid-state amplifiers was shown in Figure 1b, which was almost the same with [18]. To eliminate the influence of the amplified spontaneous emission (ASE) and accurately obtain the gain characteristics, the pump beam diameters of the 2nd and 3rd amplifiers were enlarged to about 940 μm. Moreover, in order to prevent possible parasitic laser oscillation between the 2nd and 3rd amplifiers with high gain, a 3° wedge on one of the facets of the Nd:YVO$_4$ crystal was cut for the 3rd amplifier.

The SHG was realized by a lithium borate (LBO) crystal ($\Phi = 90°$, $\theta = 0°$), with a dimension of 3 mm × 3 mm × 25 mm. To focus the output beam radius into a small point with a beam radius of

90 μm, a focusing lens with a focal length of 150 mm was employed. The LBO crystal was placed in an oven whose temperature was controlled at 160.1 °C with a precision of 0.1 °C.

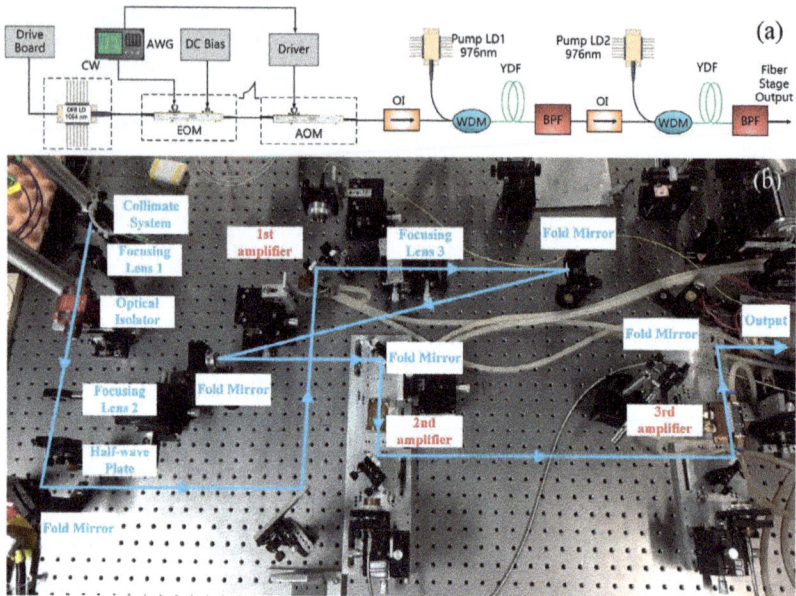

**Figure 1.** (a) Schematic of the fiber stage. OI: optical isolator; WDM: wavelength division multiplex; YDF: Ytterbium doped fiber; BPF: band pass filter (5 nm). (b) Photo of setup of solid-state laser amplifiers.

## 3. MOPA Laser with Arbitrary Pulse Shape

### 3.1. Fiber Amplifier Stage

Due to the narrow spectral linewidth of the semiconductor laser, stimulated Brillouin scattering (SBS) could happen when the pulse energy increases. We carefully increased the pump power of the 2nd fiber amplifier and observed at the output end. When the average output power reached 22 mW at a pulse repetition rate of 100 kHz with a pulse duration of 168 ns, a strong backward SBS was generated. As shown in Figure 2b, a depression appeared in the middle of the shape, and the pulse shape was also unstable (see the uploaded video from Supplementary Materials). By introducing the optical spectrum analyzer (Agilent 86142B, Palo Alto, CA, USA), the frequency shift caused by SBS was evidently observed as shown in Figure 2a. The spectrum was also unstable (see the uploaded video from Supplementary Materials).

Compared with the experimental results in [18], the SBS effect was not observed even when the output power was larger than 150 mW. The main reason is that the spectrum of directly modulated DFB laser is chirped broadened, causing a higher SBS threshold. Therefore, it is important to carefully increase the pump power during amplification in order to protect the optical components from the SBS giant pulse.

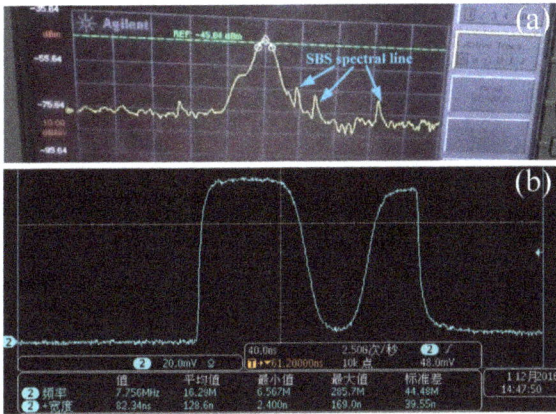

**Figure 2.** The results are measured at a pulse repetition rate of 100 kHz with a pulse duration of 168 ns. (**a**) Raman frequency shift from the stimulated Brillouin scattering (SBS) effect; (**b**) pulse distortion due to backward SBS generation.

## 3.2. Arbitrary Pulse Shape Output for the First Nd:YVO$_4$ Amplifier

In this section, we verified the feasibility of the theoretical method. By introducing three pulses and compensating the pulse distortion, the shape of a car, interestingly, was obtained. The input average power was 19.5 mW at a pulse repetition rate of 100 kHz, with a square pulse and pulse width of 168 ns.

First, the small-signal gain was calculated through our model [19]. Since the stimulated emission section $\sigma_{es}$ directly affects the gain integral, it is a key parameter for the small-signal gain. For Nd:YVO$_4$ crystal, $\sigma_{es}$ is strongly dependent on the boundary temperature [20,21]. We measured the temperature of the crystal at the pumping input end when the pump power was 56 W. The result was 120 °C, and $\sigma_{es}$ was estimated to be $12.5 \times 10^{-19}$ cm$^2$ [21–24]. Therefore, we obtained the small-signal gain of the first solid-state stage as 40 dB by slice model [19]. By applying the iteration method [17], the initial pulse gain of 26.5 dB obtained only after 25 times, as depicted by Figure 3. The pulse shape transfer functions from experiment and simulation were plotted in Figure 4a, while the output pulse shapes from experiment and simulation with the same input pulse shape were plotted in Figure 4b. The simulated output power was 7.38 W, being consistent with the experimental one of 7.32 W. Comparing the experimental and simulated results, it was concluded that the theoretical method was very effective in calculating the gain characteristics and in obtaining the pulse shape transfer function.

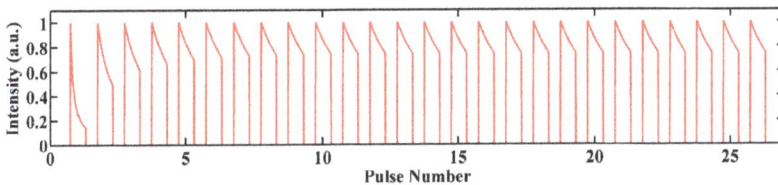

**Figure 3.** Pulse shape evolution during the iteration.

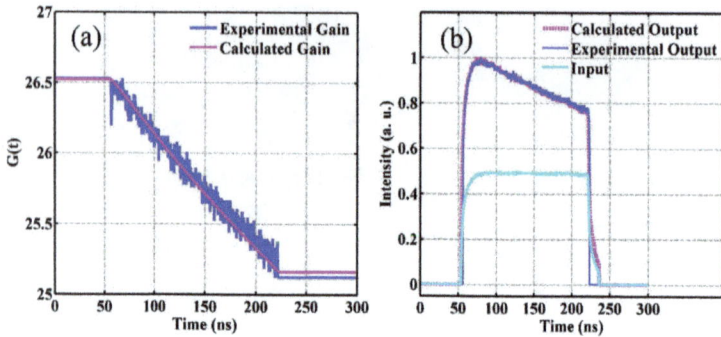

**Figure 4.** (a) Pulse shape transfer function from experiment and simulation. (b) Output pulse shape from experiment and simulation with the same input pulse shape.

To test the capability of compensating pulse distortion and obtain car-shaped pulse shapes, the input pulses were theoretically obtained by the desired output pulse shapes and $G(t)$. With the same amplifier setup, the pulse shape transfer function was the same as that in Figure 4. As shown in Figure 5a, for the input pulses, the leading edges' amplitude values were smaller than those of the trailing edges. As the gain of the leading edges were larger than that of the trailing edges, the output shapes were car-shaped, as plotted in Figure 5b. Comparing the target pulse shapes (in black dot lines) and the experimental one, they are in good accordance with each other. As for the output power of the three pulse shapes, they were all equal to 7.3 W with identical input powers of 19.5 mW.

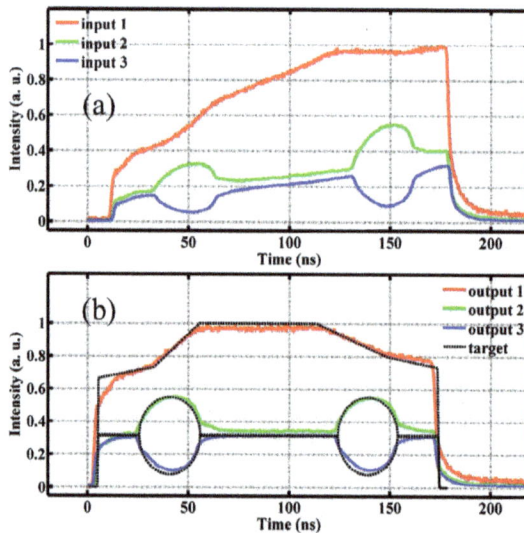

**Figure 5.** Compensation of the pulse distortion. (a) Input pulse shape (three pulses); (b) output pulse shape with amplified power.

### 3.3. Arbitrary Pulse Shape Output for the MOPA System

To further boost the output power and obtain the desired output pulse shapes, another two Nd:YVO$_4$ amplifiers were cascaded after the first Nd:YVO$_4$ amplifier. The output power of 7.3 W was scaled up to 21 W and 41.4 W by the two amplifiers, respectively. With the help of a beam propagation analyzer (Spiricon M2-200-FW-SCOR, Ophir, Israel), the beam quality was measured to be very good. The $M^2$ factor were 1.150 and 1.113, respectively, at horizontal and vertical directions. Due to the setup of the amplifiers, the ASE and parasitic oscillation was not significantly observed at the output of the whole system, both from the pulse shape and the spectrum. The optical signal-to-noise ratio (OSNR) was as high as 42 dB. The spectral line-width (3 dB) was so narrow that it exceeded the resolution of the optical spectrum analyzer (Agilent 86142B, Palo Alto, CA, USA). Many times, the resolution was measured to be less than 0.060 nm. Therefore, we concluded that the spectral linewidth was narrower than 0.06 nm.

The pulse shapes were recorded for each amplifier as shown in Figure 6. The pulse shape distorted after each amplifier resulting from different gains for the leading and trailing edge. The leading edge first consumes the inversion population resulting in a smaller gain for the falling edge and then pulse distortion.

Several desired pulse shapes were obtained after the third amplifier through the same method. The initial pulse gain for the MOPA system was calculated as 34.7 dB. As shown in Figure 7, four common pulse shapes, including multiple-step, parabolic, Gaussian, and flat-top pulses, were introduced. The corresponding input pulses are shown in the insets.

To enhance the peak power, the repetition rate was set to be 40 kHz, and the pulse duration was 40 ns. The output power was 38.3 W, corresponding to the pulse energy of 0.96 mJ, with the input power of 13.5 mW. Using the same method, square, parabolic, and Gaussian pulses were obtained for the MOPA system, as shown in Figure 8. The highest peak power of 41.6 kW, much larger than that in [9], was achieved by the Gaussian pulse.

Limited by the rising time and falling time of the arbitrary waveform generator (AWG), the leading edge and falling edge of the pulses could not be controlled very precisely, especially for the square pulses with fast leading edges and falling edges. In addition, the control of the null point of the EOM and the inadequate sampling rate of the AWG led to the small mismatch of the targeted pulse shape.

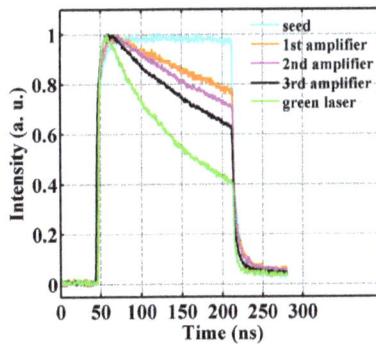

**Figure 6.** Pulse evolution for each amplifier.

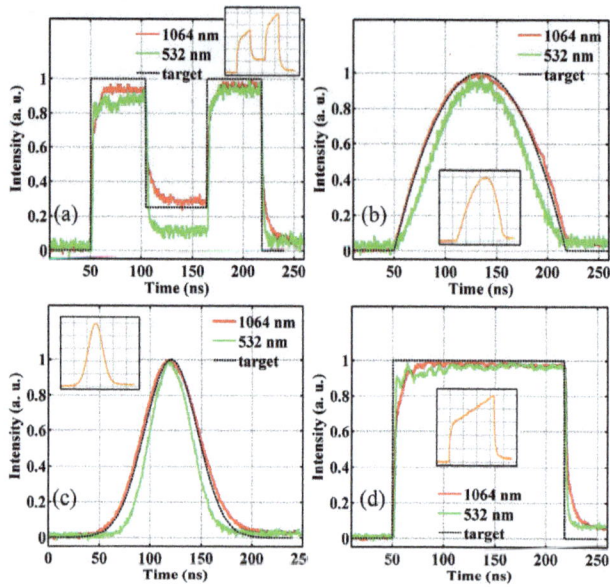

**Figure 7.** Desired pulse shapes for the 1064 nm and 532 nm at a PRF of 100 kHz. Target is for 1064 nm. (**a**) Multiple-step; (**b**) Parabolic; (**c**) Gaussian; (**d**) Square. (Input pulse shape is given in the inset).

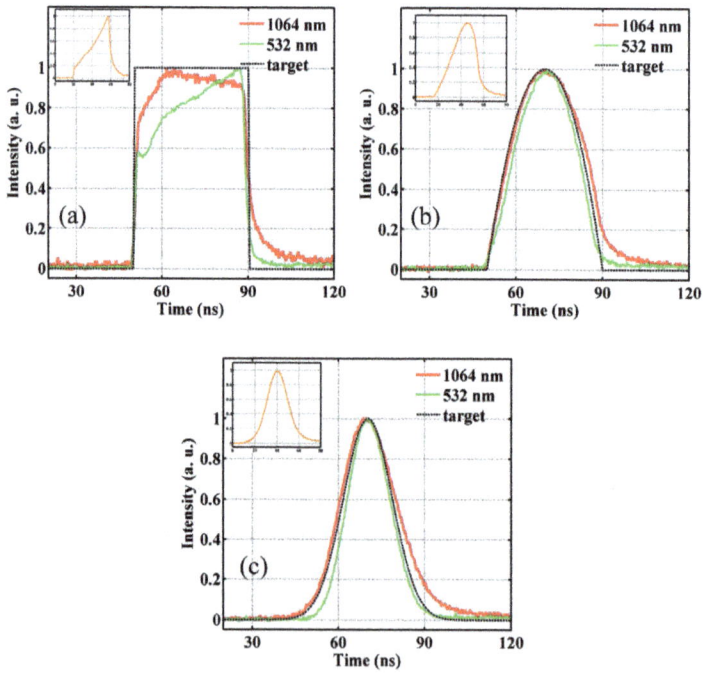

**Figure 8.** Desired pulse shapes for the 1064 nm and 532 nm at a PRF of 40 kHz. Target is for 1064 nm. (**a**) Square; (**b**) Parabolic; (**c**) Gaussian. (Input pulse shape is given in the inset).

## 4. Second Harmonic Generation

To obtain green lasers with arbitrary pulse shapes, second harmonic generation was accomplished using the output of the above-mentioned MOPA system. The temporal parameters of the fundamental laser (1064 nm) and the output power of the green laser are listed in Table 2. The pulse shapes of the green laser are demonstrated in Figures 7 and 8 with green lines. The maximum conversion efficiency of 42.6% was achieved when peak power intensity was 163 MW/cm$^2$ at a PRF of 40 kHz.

**Table 2.** Temporal parameters of 1064 nm and an output power of 532 nm.

| Pulse Shape | 100 kHz, 41.4 W | | | | 40 kHz, 38.3 W | | | |
|---|---|---|---|---|---|---|---|---|
| | Leading Edge (ns) | Falling Edge (ns) | Pulse Duration (ns) | Power of Green Laser (W) | Leading Edge (ns) | Falling Edge (ns) | Pulse Duration (ns) | Power of Green Laser (W) |
| square | 9 (1.53) [1] | 15 (5.7) | 168 (166.8) | 3.63 | 6.6 (2.4) | 8.6 (3.6) | 39 (38) | 14.5 |
| Gaussian | 45 (32.6) | 51 (37.9) | 63 (47.2) | 5.3 | 16 (11.7) | 20 (13.7) | 23 (17.5) | 16.3 |
| parabolic | 49 (57.2) | 54 (64.4) | 123.4 (96) | 4.19 | 14 (14.4) | 18 (13.2) | 29 (24.7) | 16.0 |
| multiple-step | 8 (2.0) | 12 (7.5) | 111 (-) | 4.23 | - | - | - | - |

[1] The data in the bracket is for green laser.

Comparing the output power of the green laser with the same pulse energy but with different pulse shapes in Table 2, it was deduced that a smaller pulse duration, namely a higher peak power intensity, contributes to a higher conversion efficiency from a fundamental laser to the second harmonic wave. Moreover, the conversion efficiency is not linear to the peak power intensity, leading to pulse shape distortion from the fundamental laser. The portion with a higher peak power in the fundamental pulse converts to a green laser to a much greater extent than that with a lower peak power, resulting in a pulse duration narrowing effect.

The comparison between the 1064 nm and 532 nm lasers in Figure 7a evidently shows that the conversion efficiency differs for different peak power levels of fundamental lasers. The lower peak power in the pulse is 0.25 for the fundamental laser, while it is smaller than 0.25 for the green laser. As shown in Figures 7b and 8c, the leading edge and the falling edge of the green laser become smaller than the fundamental laser after the SHG process, which results in a narrower pulse duration. The comparison between the 1064 nm and 532 nm lasers in Figures 7d and 8a indicates that pulse shapes with a uniform distribution of peak power could alleviate the narrowing effect and obtain a green laser pulse similar to the fundamental one. For the Gaussian shape, a higher peak power is helpful to keep the original fundamental pulse shape, as the pulse shape narrowing percentage in Figures 7b and 8b are 22% and 15%, respectively.

To obtain a desired pulse shape from the SHG process, the non-linear conversion efficiency must be taken into account. For example, to obtain a multiple-step green pulse just like in Figure 7a (the lower peak power is 1/4 of the higher peak power for green laser), the lower peak power of the fundamental laser must be larger than one-fourth of the higher peak power.

## 5. Conclusions

In summary, desired pulse shapes were demonstrated in a hybrid MOPA system mainly made up of three Nd:YVO$_4$ amplifiers. By investigating the amplifier dynamics, the pulse shape transfer function of an end-pumped Nd:YVO$_4$ amplifier was obtained. The numerical simulation for the first amplifier indicates excellent agreement with the experimental results. Interesting pulse shapes, such as that of a car, were obtained through the theoretical method, showing the effectiveness of the theoretical method in achieving arbitrary pulse shapes. Furthermore, several pulse shapes were obtained through the same method for the MOPA system made up of three amplifiers. A high peak power level of 41.6 kW and a narrow linewidth less than 0.06 nm were achieved, with good beam quality. A green laser was generated by this source, with a maximum conversion efficiency of 42.6%. Compared with

*Appl. Sci.* **2017**, *7*, 892

the input pulse of SHG, the pulse shape of the green laser was distorted due to the nonlinear effect, especially for the ones with a slow leading edge and a falling edge. The solid-state MOPA system with arbitrary waveforms, high peak power levels, and narrow spectral linewidths shows great potential in precision manufacturing, optical sensing, and Lidar applications.

**Supplementary Materials:** Supplementary materials are available online at Zenodo DOI: 10.5281/zenodo.854840 (https://zenodo.org/record/854840#.WaefZ4SGMcY). Video S1: SBS pulse distortion_40kHz_40ns. Video S2: SBS pulse distortion_100kHz_168ns. Video S3: SBS_freqency shift_100 kHz_168ns.

**Acknowledgments:** National Key Research and Development Program of China (Grant No. 2017YFB1104500); National Natural Science Foundation of China (61475083).

**Author Contributions:** Mingming Nie conceived and designed the experiments; Qiang Liu provided guidance and funding; Mingming Nie performed the experiments; Encai Ji and Xing Fu analyzed the data; Xuezhe Cao conducted the simulation; Mingming Nie wrote the manuscript and created the figures. All authors have reviewed the manuscript.

**Conflicts of Interest:** The authors declare no conflict of interest.

## References

1.  Biswas, A.; Hemmati, H.; Lesh, J.R. High data rate laser transmission for free space laser communications. *SPIE Rev.* **1999**, *3615*, 269–277.
2.  Abshire, J.B.; Collatz, G.J.; Sun, X.; Riris, H.; Andrews, A.E.; Krainak, M. Laser sounder technique for remotely measuring atmospheric $CO_2$ concentrations. In Proceedings of the American Geophysical Union, Fall Meeting, San Francisco, CA, USA, 10–14 December 2001.
3.  Kelly, D.; Young, C.Y.; Andrews, L.C. Temporal broadening of ultrashort space-time Gaussian pulses with applications in laser satellite communication. *SPIE Rev.* **1998**, *3266*, 231–240.
4.  Hemmati, H.; Wright, M.; Esproles, C. High efficiency pulsed laser transmitters for deep space communications. *SPIE Rev.* **2000**, *3932*, 188–195.
5.  Wan, P.; Liu, J.; Yang, L.; Amzajerdian, F. Pulse shaping fiber laser at 1.5 μm. *Appl. Opt.* **2012**, *51*, 214–219. [CrossRef] [PubMed]
6.  Siegman, A.E. *Lasers*; University Science Books: Sausalito, CA, USA, 1986.
7.  Vu, K.T.; Malinowski, A.; Richardson, D.J.; Ghiringhelli, F.; Hickey, L.M.B.; Zervas, M.N. Adaptive pulse shape control in a diode-seeded nanosecond fiber MOPA system. *Opt. Express* **2006**, *14*, 10996–11001. [CrossRef] [PubMed]
8.  Schimpf, D.N.; Ruchert, C.; Nodop, D.; Limpert, J.; Tuennermann, A.; Salin, F. Compensation of pulse-distortion in saturated laser amplifiers. *Opt. Express* **2008**, *16*, 17637–17646. [CrossRef] [PubMed]
9.  Malinowski, A.; Vu, K.T.; Chen, K.K.; Nilsson, J.; Jeong, Y.; Alam, S.; Lin, D.; Richardson, D.J. High power pulsed fiber MOPA system incorporating electro-optic modulator based adaptive pulse shaping. *Opt. Express* **2009**, *17*, 20927–20937. [CrossRef] [PubMed]
10. Sobon, G.; Kaczmarekm, P.; Antonczak, A.A.; Sotor, J.; Waz, A.; Abramski, K.M. Pulsed dual-stage fiber MOPA source operating at 1550 nm with arbitrarily shaped output pulses. *Appl. Phys. B* **2011**, *105*, 721–727. [CrossRef]
11. Heidt, A.M.; Li, Z.; Richardson, D.J. High Power Diode-Seeded Fiber Amplifiers at 2 μm—From Architectures to Applications. *IEEE J. Sel. Top. Quantum Electron.* **2014**, *20*, 525–536. [CrossRef]
12. Smith, R.G. Optical power handling capacity of low loss optical fibers as determined by stimulated Raman and Brillouin scattering. *Appl. Opt.* **1972**, *11*, 2489–2494. [CrossRef] [PubMed]
13. Shaw, M.; Williams, W.; House, R.; Haynam, C. Laser Performance Operations Model (LPOM). In *Inertial Confinement Fusion Semiannual Report*; Lawrence Livermore National Laboratory: Livermore, CA, USA, 2004.
14. Morice, O. Miró: Complete modeling and software for pulse amplification and propagation in high-power laser systems. *Opt. Eng.* **2003**, *42*, 1530–1541. [CrossRef]
15. Siegman, A.E. Design considerations for laser pulse amplifiers. *J. Appl. Phys.* **1964**, *35*, 460. [CrossRef]
16. Xu, L.; Yu, L.; Chu, Y.; Gan, Z.; Liang, X. Temporal compensation method of pulse distortion in saturated laser amplifiers. *Appl. Opt.* **2015**, *54*, 357–362. [CrossRef]
17. Nie, M.; Cao, X.; Liu, Q.; Ji, E.; Fu, X. 100 μJ pulse energy in burst-mode-operated hybrid fiber-bulk amplifier system with envelope shaping. *Opt. Express* **2017**, *25*, 13557–13566. [CrossRef] [PubMed]

18.  Nie, M.; Liu, Q.; Ji, E.; Gong, M. High peak power hybrid MOPA laser with tunable pulse repetition frequency and pulse duration. *Appl. Opt.* **2017**, *56*, 3457–3461. [CrossRef] [PubMed]

19.  Nie, M.; Liu, Q.; Ji, E.; Cao, X.; Fu, X.; Gong, M. Design of High-Gain Single-Stage and Single-Pass Nd:YVO$_4$ Amplifier Pumped by Fiber-Coupled Laser Diodes: Simulation and Experiment. *IEEE J. Quantum Electron.* **2016**, *52*, 1–10. [CrossRef]

20.  Delen, X.; Balembois, F.; Georges, P. Temperature dependence of the emission cross section of Nd:YVO$_4$ around 1064 nm and consequences on laser operation. *J. Opt. Soc. Am. B* **2011**, *28*, 972–976. [CrossRef]

21.  Sato, Y.; Taira, T. Temperature dependencies of stimulated emission cross section for Nd-doped solid-state laser materials. *Opt. Mater. Express* **2012**, *2*, 1076–1087. [CrossRef]

22.  Nie, M.; Liu, Q.; Ji, E.; Gong, M. Gain change by adjusting the pumping wavelength in an end-pumped Nd:YVO$_4$ amplifier. *Appl. Opt.* **2015**, *54*, 8383–8387. [CrossRef] [PubMed]

23.  Krishnan, G.; Bidin, N. Determination of a stimulated emission cross section gradient on the basis of the performance of a diode-end-pumped NdYVO$_4$ lase. *Laser Phys. Lett.* **2013**, *10*, 1–5. [CrossRef]

24.  Peterson, R.; Jenssen, H.; Cassanho, A. Investigation of the Spectroscopic Properties of Nd:YVO$_4$. In Proceedings of the Advanced Solid-State Lasers (Optical Society of America, 2002), Québec City, QC, Canada, 3 February 2002; pp. 294–298.

*applied*
*sciences*

MDPI

*Article*

# Non-Pulse-Leakage 100-kHz Level, High Beam Quality Industrial Grade Nd:YVO$_4$ Picosecond Amplifier

Zhenao Bai [1,2,3,†], Zhenxu Bai [4,5,†], Zhijun Kang [1,2], Fuqiang Lian [1,2], Weiran Lin [1,2] and Zhongwei Fan [1,2,3,6,*]

[1]  Academy of Opto-Electronics, Chinese Academy of Sciences, Beijing 100094, China; baizhenao@hotmail.com (Z.B.); kzjun1221@126.com (Z.K.); aoefiberlaser@126.com (F.L.); linweiran@aoe.ac.cn (W.L.)
[2]  National Engineering Research Center for DPSSL, Beijing 102211, China
[3]  Sino-HG Applied Laser Technology Institute Company, Ltd., Tianjin 300304, China
[4]  National Key Laboratory of Science and Technology on Tunable Laser, Harbin Institute of Technology, Harbin 150001, China; baizhenxu@hotmail.com
[5]  MQ Photonics Research Centre, Department of Physics and Astronomy, Macquarie University, Sydney, NSW 2109, Australia
[6]  University of Chinese Academy of Sciences, Beijing 100049, China
[*]  Correspondence: fanzhongwei@aoe.ac.cn; Tel.: +86-10-8217-8609
[†]  These authors contributed equally to this work.

Academic Editor: Federico Pirzio
Received: 27 April 2017; Accepted: 26 May 2017; Published: 14 June 2017

**Abstract:** A non-pulse-leakage optical fiber pumped 100-kHz level high beam quality Nd:YVO$_4$ picosecond amplifier has been developed. An 80 MHz, 11.5 ps mode-locked picosecond laser is used as the seed with single pulse energy of 1 nJ. By harnessing the double β-BaB$_2$O$_4$ (BBO) crystal Pockels cells in both the pulse picker and regenerative amplifier, the seed pulse leakage of the output is suppressed effectively with an adjustable repetition rate from 200 to 500 kHz. Through one stage traveling-wave amplifier, a maximum output power of 24.5 W is generated corresponding to the injected regenerative amplified power of 9.73 W at 500 kHz. The output pulse duration is 16.9 ps, and the beam quality factor $M^2$ is measured to be 1.25 with near-field roundness higher than 99% at the full output power.

**Keywords:** regenerative amplifier; double-crystal Pockels cell; 100-kHz; non-pulse-leakage

## 1. Introduction

Over the past decade, there has been significant development of high power ultrashort pulse lasers with high repetition rate [1–3], with applications including nonlinear optical frequency conversion, precise material processing, satellite ranging, and high-intensity physics [4–7]. Specifically, in ultrashort lasers in the order of picoseconds, their pulse width is smaller than the electron-phonon coupling time of most materials, heat conduction can be decreased substantially during the process of interaction. This decrease is beneficial to fine processing, biological medicine, and historical relic cleaning [8–11]. Currently, mode-locking is the main approach to obtain the pulse width around ~10 ps. However, the pulse energy of the directly generated mode-locked lasers is very low such that it restricts their applications. To solve this problem, different types of laser amplifiers are utilized, in which a regenerative amplifier is considered to be the most effective method for strong amplification of the mode-locked pulses [1,2,12,13]. In most cases, increasing the repetition rate and power can improve the efficiency of laser processing and laser ranging. Therefore, it is not only necessary to increase the single pulse energy but also the repetition rate of the current regenerative amplified lasers. Determined by the parameters of Pockels cell and regenerative cavity, repetition rates from hundreds to mega-Hz have been obtained with output power up to 100 watts. For instance, Nd:YVO$_4$ picosecond regenerative amplifiers with a repetition rate up to 850 kHz have been demonstrated [14,15]; other broadband gain materials such as Yb:CaF$_2$, Yb:CAlGO, Yb:YVO$_4$, and Nd:LuVO$_4$ show up to 1.43 MHz repetition rate [16–18]. Currently, Bergmann et al. [19] reported the record high repetition rate of 2 MHz for a picosecond regenerative amplifier in Yb:YAG. In general, when the repetition rate is up to 100 kHz, it is very difficult for the Pockels cell to realize the fast and complete switching due to the limitations of high voltage driving supply. This will lead to the leakage of the directly amplified seed pulses in the output that not only introduce optical noise but also increase unnecessary pump power consumption. Especially after traveling-wave amplification, the existence of pulse leakage will further reduce the master-slave pulse ratio of the output and influence the performance of the laser system in applications. For example, in nonmetallic brittle materials processing (viz. artificial crystal, ceramics, and solar panels), the amplified leaking pulses easily cause breakage of the samples because of the excessive heat load; meanwhile, the presence of leaked pulses also results in the reduction of accuracy in laser satellite ranging and laser communication due to the introduction of stray signals. Therefore, it is necessary to eliminate the pulse leakage of regenerative amplifiers used in some specific industrial areas.

Nd:YVO$_4$ crystals as a gain material with sufficient gain bandwidth and a high emission cross section are widely adopted in end-pumping picosecond regenerative amplifiers. Nowadays, fiber coupled high-power laser diode (LD) are an appropriate pumping source for the regenerative amplifier due to their highly compact size and relatively low cost. The pumping of Nd:YVO$_4$ with 888 nm LD induces less thermal stress gradient along the crystal compared with 808 nm that makes it possible for long time operation of the laser with higher stability [20,21]. Table 1 presents the physical parameters of Nd:YVO$_4$ crystal pumped by 808 nm and 888 nm wavelengths [22,23]. As the data show, the lower fractional thermal loading in 888 nm pump compared with 808 nm wavelength results in the heat reduction and optical efficiency growth. Although the absorption coefficient of 808 nm is higher compared to 888 nm wavelength, it shows different values between the two crystallographic axes that result in the varying and irreproducible absorption if without the polarization controlling of pump light. The lower pump absorption efficiency can be optimized by using a longer crystal for high output power that allows the heat load to be spread in a larger volume, thus minimizing stress and thermal gradient [24]. Additionally, the 888 nm wavelength shows a higher absorption band width than 808 nm, which decreases the sensitivity to the shift in the pump wavelength.

**Table 1.** Comparison of Nd:YVO$_4$ pumped by 808 nm and 888 nm.

| Wavelength (nm) | Output Wavelength (nm) | Fractional Thermal Loading | Absorption Coefficient (cm$^{-1}$) | Absorption Band Width (nm) |
|---|---|---|---|---|
| 808 | 1064 | 0.241 | 10 (*a*-axis) 37 (*c*-axis) | 0.8 |
| 888 | 1064 | 0.173 | 1.5 (*a*- & *c*-axis) | 3 |

In this study, we demonstrate a highly stable and high power industrial grade picosecond Nd:YVO$_4$ amplifier with the adjustable repetition rate from 200 to 500 kHz. A combination of an 888 nm optical fiber pumped Nd:YVO$_4$ regenerative and a traveling-wave amplifiers have been utilized. To eliminate the leakage of the regenerative amplifier in high repetition rate operation, double BBO crystal Pockels cells are adopted as one switch to improve the responsivity of the switching. The maximum output power of 24.5 W is obtained at 500 kHz with the pulse width of 16.9 ps, corresponding to the pulse energy and the peak power of 0.05 mJ and 2.9 MW, respectively. The beam quality factor $M^2$ is measured to be 1.25 at the full power operation. When operated at 200 kHz, the maximum pulse energy of 0.11 mJ and the maximum peak power of 6.5 MW is obtained. Our measurements show that this method effectively realizes the suppression of leakage and improves the purity of the regenerative amplified pulses.

## 2. Experimental Principle and Setup

Q-switch is the core component of the regenerative amplifier that is used in the pulse picker for mode-locked sequence. To date, both acousto- and electro-optic switches have been used to realize the pulse selection function in regenerative amplifiers [19,25,26]. However, due to the slow switching speed (with rise time up to 100 nanoseconds) and limited diffraction efficiency (usually less than 90%), the acousto-optic switch is difficult to generate regenerative pulses without the leakage of mode-locked seed. In contrast to the acousto-optic switch, the electro-optic switch has faster switching speed (rise time <10 ns) and excellent switching effect, which is widely adopted in regenerative amplifiers. Nowadays, the commonly used electro-optic crystals include β-BaB$_2$O$_4$ (BBO), KH$_2$PO$_4$ (KDP), KD$_2$PO$_4$ (KD * P), RbTiOPO$_4$ (RTP), and LiNbO$_3$ (Lithium niobate) etc. To meet different experimental conditions (such as half- or quarter-wavelength voltage, acceptable input beam diameter, and available crystal size), both longitudinal and transverse electrode configuration Pockels cells are manufactured. However, so far there is not one kind of electro-optic crystal or configuration the can provide a general solution under different operating wavelengths, repetition rates, spot diameters, and output powers. At present, BBO crystal Pockels cells with minimal piezoelectric ringing and low acoustic noise, have emerged in high repetition rate (>100 kHz) pulse picking applications [14–17,19]. Nevertheless, with the increase of repetition rate, it is difficult to realize fast switching within the effective duty cycle due to the long fall time of the high voltage.

To solve the problem, double BBO crystal Pockels cell is adopted in our experiment to reduce the load of the high voltage driving supply by reducing the quarter-wavelength voltage. The value of the quarter-wavelength voltage $V_{\lambda/4}$ of BBO crystal is given by [27,28]:

$$V_{\lambda/4} = \frac{\lambda d}{4n_0^3 d_{22} L} \tag{1}$$

where laser wavelength λ = 1064 nm, the crystal section thickness $d$ = 3 mm, the refractive index of the crystal $n_0$ = 1.66, the effective electro-optic coefficient $d_{22}$ = 2.2 pm/V, and the crystal length $L$ = 40 mm. The calculated value of the quarter-wavelength voltage is just 2000 V, which is almost half that of the previous reports. The lower quarter-wavelength voltage of Pockels cell means faster switching speed, and makes it possible for higher repetition rate operation.

The experimental setup of the picosecond amplifier is shown in Figure 1, which consists of a picosecond seed source, two optical isolators, a BBO pulse picker, a regenerative amplifier, and one stage traveling-wave amplifier.

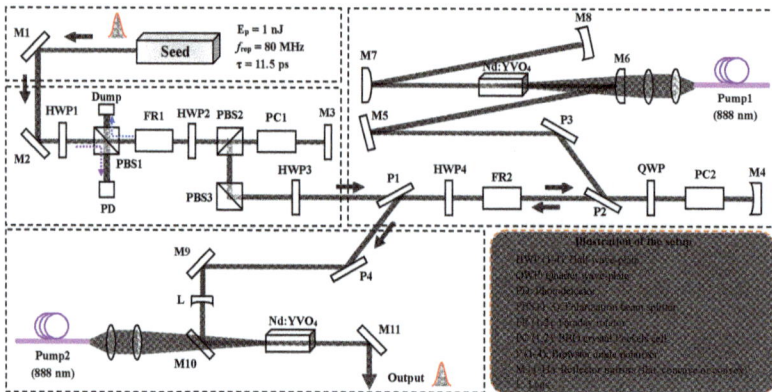

**Figure 1.** Diagram of the non-leakage 100-kHz Nd:YVO$_4$ picosecond regenerative amplifier. The inset is the illustration of the setup.

In the present system, the amplifier is seeded with a self-developed semiconductor saturable absorption mirror (SESAM) mode-locked oscillator capable of generating 11.5 ps pulses at a repetition rate of 80 MHz with a single pulse energy of 1 nJ. Picosecond pulse sequence from the seed is injected into the pulse picker through two high-reflection mirrors, M1 and M2. The function of the pulse picker is to select the desired repetition rate for regenerative amplifier by eliminating the pulse leakage of the seed sequence. The half-wave plate HWP1 is set to let most of the seed light to $p$-polarization and the rest $s$-polarization reflected into the photodetector (PD) as feedback signal. After passing through the first stage optical isolator composed of a polarization beam splitter PBS1, a Faraday rotator FR1, and a HWP2, the seed enters into the double BBO crystal Pockels cell PC1 (PCB3S-1342; EKSMA Optics, Vilnius, Lithuania). The size of the two BBO crystals are $3 \times 3 \times 20$ mm$^3$. The distance between the PC1 and M3 is 0.8 m, corresponding to a round-trip time about 5.3 ns. Without voltage of the PC1, the reflected seed from the M3 will pass through the PBS2 and FR1, then enter into the dump. While with a quarter-wavelength voltage of the PC1, the pulse sequence with a specified range of repetition rate can be selected with polarization changed to $s$-polarization after double-passing through PC1 and then output by the PBS2. The switching time of the two Pockels cells in our setup is adjustable from 25 to 200 ns with a rising and falling time about 5 ns. The switch time of the PC1 is set to be 20 ns. With the polarization changed to $p$-polarization by HWP3, the selected pulses enter into the regenerative amplifier. The second optical isolator formed by a Brewster angle polarizer P1, HWP4, and FR2 is placed in front of the regenerative amplifier which is used to output the regenerative amplified pulse.

The laser pulse with $p$-polarization is injected into the regenerative amplifier by the P2. The length of the whole cavity is 1.8 m, corresponding to a cavity round-trip time of 12 ns. The gain medium is 0.5 at. % doped Nd:YVO$_4$ with a size of $3 \times 3 \times 20$ mm$^3$ and double-end-wedged cut at 2°. Both sides of the Nd:YVO$_4$ have anti-reflective (AR) coating at the 1064 nm and 888 nm wavelength. A 50 W, 888 nm fiber-coupled laser diode with a numerical aperture of 0.22 and a diameter of 400 μm is used as the pump source. The coupling ratio of the pump beam is 1:3 into the Nd:YVO$_4$ crystal. M4 and M8 are two concave mirrors with R = −2000 mm; M4 and M7 are two 1064 nm high-reflection (HR) convex mirrors with R = 1500 mm in which M4 is also 888 nm AR coated. Reflected by an end mirror M4, the seed beam double passes the quarter-wave plate (HWP) and reflects through P2 with $s$-polarization. Then the PC2 inside the cavity is switched on with a quarter-wavelength voltage and the seed laser continues to make round trips in the regenerative cavity with $s$-polarization until the PC2 is switched off. The regenerative amplified pulses are reflected by P1 with $s$-polarization, and then enter into the traveling-wave amplifier.

To further amplify the output power, one stage traveling-wave Nd:YVO$_4$ amplifier is adopted. A Nd:YVO$_4$ crystal with the dimensions of $4 \times 4 \times 30$ mm$^3$ is selected as gain medium with 0.5 at. % Nd$^{3+}$-doped and at a 2° angle. The pumping source of the traveling-wave amplifier has the same parameters as the regenerative amplifier with a coupling ratio of 1:4. L is a concave lens with f = −360 mm used to compensate the thermal lens effect of the Nd:YVO$_4$ crystal. M10 and M11 are two 888 nm AR & 1064 nm HR coated 45° plane mirror. The single-pass amplified pulse is output from the M11.

## 3. Experimental Results and Discussion

In our experiment, we tested the leakage of the regenerative amplifier by monitoring the output power with the PC2 turning on and off. At the maximum pump power of 50 W, the output power of the regenerative amplifier increases with the Pockels cells switching time, however, the leakage appears when the PC2 switching time more than 69 ns. For example, with the switching time of 79 ns, the leakage reached 2 W when the maximum regenerative amplified power was about 18 W at 500 kHz. Accordingly, in order to obtain maximum power regenerative output without leakage, the switching time of Pockels cells is set to be 68 ns, corresponding to the five round trips in the regenerative cavity. The pump power of the regenerative amplifier is set to be 50 W because higher power may result in the self-excited oscillation between the M4 and M8. Next, improvement of the leakage and self-excited oscillation threshold to increase the output power of regenerative amplifier will be studied, including the optimization of cavity design and Pockels cell structure (e.g., employing bidirectional voltage switching power supply to achieve a faster and more thorough switch [29]).

The injected pulse energy into the regenerative amplifier was about 1 nJ, and the output pulse energy saturated with the increase number of round trips in the cavity. Figure 2a,b illustrates the dependence of the output power and pulse energy of the regenerative and final amplifier output on the repetition rate from 200 to 500 kHz, respectively. At the same pump power, the out power of the regenerative amplifier increases with the repetition rate, while the single pulse energy decreases, which is shown in Figure 2a. At 500 kHz, maximum regenerative amplified power of 9.73 W was obtained corresponding to the pulse energy of 0.02 mJ. After passing through the traveling-wave amplifier, the maximum output power of the laser system was 24.5 W at 500 kHz, corresponding to the peak power of 2.9 MW; while the maximum output pulse energy obtained was 0.11 mJ at 200 kHz, corresponding to the peak power of 6.5 MW. No leakage occurred when the repetition rate continuously adjusting from 200 to 500 kHz with the measured root-mean-square-error (RMSE) of output power less than 0.04%.

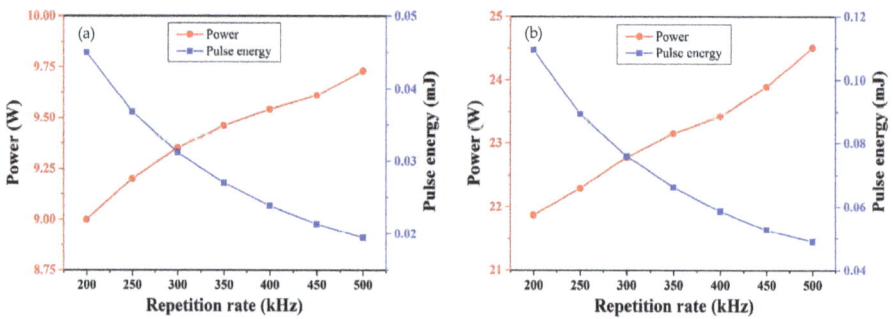

**Figure 2.** Average output power and pulse energy of the amplifier in dependence on the repetition rate of (a) regenerative amplifier output; and (b) final amplifier output.

The amplified pulses are monitored using a photodiode. At 500 kHz, oscilloscope trace of the output pulse train with 4 µs/div and single pulse with 10 ns/div are illustrated in Figure 3, respectively. We can observe that the amplifier generates very clean output pulses without any noticeable pulse fluctuation or stray signal.

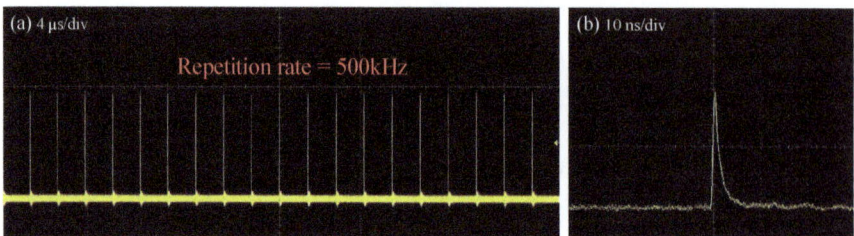

**Figure 3.** Oscilloscope trace of the regenerative amplified pulse train (a) 4 µs/div; and (b) 10 ns/div.

Figure 4a shows the autocorrelation trace of the amplified pulses with the measured width of 16.9 ps at a repetition rate of 500 Hz. The pulse width of the output was slightly broadened compared with that of the seed pulses 11.5 ps, which is mainly caused by the gain narrowing [30,31]. The measured beam quality factor $M^2$ and near-field beam intensity profile are shown in Figure 4b.

The $M^2$ measured were 1.27 and 1.22 for the horizontal and vertical axes of the output, respectively. The roundness of the near-field intensity distribution is higher than 99%.

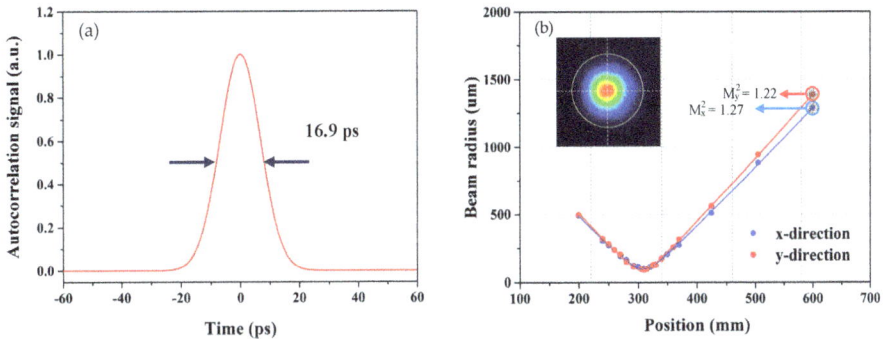

**Figure 4.** Measured (**a**) autocorrelation trace (Gaussian fitting) for the amplified pulse; and (**b**) beam quality factor $M^2$ (insert: near-field beam intensity distribution).

## 4. Conclusions

In conclusion, we have demonstrated a non-pulse-leakage high power industrial grade Nd:YVO$_4$ picosecond amplifier. Two double BBO crystal Pockels cells were used in the pulse picker and regenerative amplifier, respectively. This design successfully suppresses the appetence of pulse leakage in the regenerative amplifier that improved the purity of the output pulses, which makes it possible for the regenerative amplifier to operate stably at the repetition rate of 100-kHz level with lower high voltage driving supply. A maximum power of 24.5 W was obtained at 500 kHz, and a maximum single pulse energy 0.11 mJ was obtained at 200 kHz. High beam quality output is obtained with $M^2$ about 1.25 and the roundness of near-field distribution up to 99%. The output pulse width measured to be 16.9 ps with the RMSE of the output is less than 0.04%. Our work offers a new approach to generate high quality (in both time domain and space domain) 100-kHz level repetition rate continuous adjustable picosecond regenerative amplified output. This solution can be made possible with high efficiency and high precision processing for nonmetallic brittle materials, as well as low noise space communication.

**Acknowledgments:** This work was supported by the Development of High Power Nanosecond Laser & Precision Detecting Instrument Foundation (Grant No. ZDYZ2013-2) and China Innovative Talent Promotion Plans for Innovation Team in Priority Fields (Grant No. 2014RA4051). We also acknowledge Mojtaba Moshkani (Macquarie University) for his helpful comments.

**Author Contributions:** We confirm that all authors contributed substantially to the reported work. Zhenao Bai was the originator of the idea of this study and conceived most of the experiments; Zhenxu Bai performed the experiments and wrote the manuscript under the supervision of Zhenao Bai and Zhongwei Fan; Zhijun Kang, Fuqiang Lian, and Weiran Lin participated in the research design and analyzed the data; Zhongwei Fan supervised the research and provided the facilities. All the authors discussed and interpreted the results. All the authors read the final manuscript.

**Conflicts of Interest:** The authors declare no conflict of interest.

## References

1.  Siebold, M.; Hornung, M.; Hein, J.; Paunescu, G.; Sauerbrey, R.; Bergmann, T.; Hollemann, G. A high-average-power diode-pumped Nd:YVO$_4$ regenerative laser amplifier for picosecond-pulses. *Appl. Phys. B* **2004**, *78*, 287–290. [CrossRef]
2.  Metzger, T.; Schwarz, A.; Teisset, C.Y.; Sutter, D.; Killi, A.; Kienberger, R.; Krausz, F. High-repetition-rate picosecond pump laser based on a Yb:YAG disk amplifier for optical parametric amplification. *Opt. Lett.* **2009**, *34*, 2123–2125. [CrossRef] [PubMed]

3.   Röser, F.; Eidam, T.; Rothhardt, J.; Schmidt, O.; Schimpf, D.N.; Limpert, J.; Tünnermann, A. Millijoule pulse energy high repetition rate femtosecond fiber chirped-pulse amplification system. *Opt. Lett.* **2007**, *32*, 3495–3497. [CrossRef] [PubMed]
4.   Zhang, Z.; Zhang, H.; Wu, Z.; Chen, J.; Li, P.; Yang, F. kHz repetition Satellite Laser Ranging system with high precision and measuring results. *Chin. Sci. Bull.* **2011**, *56*, 1177–1183. [CrossRef]
5.   Bai, Z.; Cui, C.; Liu, Z.; Yuan, H.; Wang, H.; Wang, Y.; Lu, Z. Drilling study on Cu, Mo, W and Ti by using SBS pulse compressed steep leading edge hundred picoseconds laser. *Optik* **2016**, *127*, 11156–11160. [CrossRef]
6.   Surmeneva, M.; Nikityuk, P.; Hans, M.; Surmenev, R. Deposition of Ultrathin Nano-Hydroxyapatite Films on Laser Micro-Textured Titanium Surfaces to Prepare a Multiscale Surface Topography for Improved Surface Wettability/Energy. *Materials* **2016**, *9*, 862. [CrossRef]
7.   Je, G.; Malka, D.; Kim, H.; Hong, S.; Shin, B. A study on micro hydroforming using shock wave of 355 nm UV-pulsed laser. *Appl. Surf. Sci.* **2017**, in press. [CrossRef]
8.   Niemz, M.H.; Klancnik, E.G.; Bille, J.F. Plasma-mediated ablation of corneal tissue at 1053 nm using a Nd:YLF oscillator/regenerative amplifier laser. *Lasers Surg. Med.* **1991**, *11*, 426–431. [CrossRef] [PubMed]
9.   Chichkov, B.N.; Momma, C.; Nolte, S.; Von Alvensleben, F.; Tünnermann, A. Femtosecond, picosecond and nanosecond laser ablation of solids. *Appl. Phys. A* **1996**, *63*, 109–115. [CrossRef]
10.  Biswas, S.; Karthikeyan, A.; Kietzig, A.M. Effect of Repetition Rate on Femtosecond Laser-Induced Homogenous Microstructures. *Materials* **2016**, *9*, 1023. [CrossRef]
11.  Elnaggar, A.; Fitzsimons, P.; Lama, A.; Fletcher, Y.; Antunes, P.; Watkins, K.G. Feasibility of ultrafast picosecond laser cleaning of soiling on historical leather buckles. *Herit. Sci.* **2016**, *4*, 30. [CrossRef]
12.  Dorrer, C.; Consentino, A.; Irwin, D.; Qiao, J.; Zuegel, J.D. OPCPA front end and contrast optimization for the OMEGA EP kilojoule, picosecond laser. *J. Opt.* **2015**, *17*, 094007. [CrossRef]
13.  Chen, Y.; Liu, K.; Yang, J.; Yang, F.; Gao, H.W.; Zong, N.; Yuan, L.; Lin, Y.Y.; Liu, Z.; Peng, Q.J.; et al. 8.2 mJ, 324 MW, 5 kHz picosecond MOPA system based on Nd: YAG slab amplifiers. *J. Opt.* **2016**, *18*, 075503. [CrossRef]
14.  Lührmann, M.; Harth, F.; Theobald, C.; Ulm, T.; Knappe, R.; Nebel, A.; Klehr, A.; Erbert, G.; L'huillier, J. High average power Nd:YVO₄ regenerative amplifier seeded by a gain switched diode laser. *Proc. SPIE* **2011**, *7912*, 791210.
15.  Bai, Z.; Fan, Z.; Lian, F.; Tan, T.; Bai, Z.; Yang, C.; Kang, Z.; Liu, C. High power 888 nm optical fiber end-pumped Nd: YVO₄ picosecond regenerative amplifier at hundreds kHz. *Proc. SPIE* **2016**, *10152*, 101520S.
16.  Caracciolo, E.; Kemnitzer, M.; Guandalini, A.; Pirzio, F.; Agnesi, A.; der Au, J.A. High pulse energy multiwatt Yb:CaAlGdO₄ and Yb:CaF₂ regenerative amplifiers. *Opt. Express* **2014**, *22*, 19912–19918. [CrossRef] [PubMed]
17.  Rudenkov, A.; Kisel, V.; Matrosov, V.; Kuleshov, N. 200 kHz 5.5 W Yb³⁺:YVO₄-based chirped-pulse regenerative amplifier. *Opt. Lett.* **2015**, *40*, 3352–3355. [CrossRef] [PubMed]
18.  Gao, P.; Lin, H.; Li, J.; Guo, J.; Yu, H.; Zhang, H.; Liang, X. Megahertz-level, high-power picosecond Nd: LuVO₄ regenerative amplifier free of period doubling. *Opt. Express* **2016**, *24*, 13963–13970. [CrossRef] [PubMed]
19.  Bergmann, F.; Siebold, M.; Loeser, M.; Röser, F.; Albach, D.; Schramm, U. MHz Repetion Rate Yb: YAG and Yb: CaF₂ Regenerative Picosecond Laser Amplifiers with a BBO Pockels Cell. *Appl. Sci.* **2015**, *5*, 761–769. [CrossRef]
20.  McDonagh, L.; Wallenstein, R.; Nebel, A. 111 W, 110 MHz repetition-rate, passively mode-locked TEM₀₀ Nd: YVO₄ master oscillator power amplifier pumped at 888 nm. *Opt. Lett.* **2007**, *32*, 1259–1261. [CrossRef] [PubMed]
21.  Bai, Z.A.; Fan, Z.W.; Bai, Z.X.; Lian, F.Q.; Kang, Z.J.; Lin, W.R. Optical fiber pumped high repetition rate and high power Nd: YVO₄ picosecond regenerative amplifier. *Appl. Sci.* **2015**, *5*, 359–366. [CrossRef]
22.  McDonagh, L.; Wallenstein, R.; Knappe, R.; Nebel, A. High-efficiency 60 W TEM₀₀ Nd: YVO₄ oscillator pumped at 888 nm. *Opt. Lett.* **2006**, *31*, 3297–3299. [CrossRef] [PubMed]
23.  Bai, Z.A.; Fan, Z.W.; Lian, F.Q.; Bai, Z.X.; Kan, Z.J.; Zhang, J. 20 W passive mode-locked picosecond oscillator. *Proc. SPIE* **2014**, *9281*, 92812M.
24.  McDonagh, L.; Knappe, R.; Nebel, A.; Wallenstein, R. 888 nm pumping of Nd: YVO₄ for high-power high-efficiency TEM₀₀ lasers. *Proc. SPIE* **2017**, *6451*, 64510F.

25. Delaigue, M.; Manek-Hönninger, I.; Salin, F.; Hönninger, C.; Rigail, P.; Courjaud, A.; Mottay, E. 300 kHz femtosecond Yb: KGW regenerative amplifier using an acousto–optic Q-switch. *Appl. Phys. B* **2006**, *84*, 375–378. [CrossRef]

26. Liu, J.; Wang, W.; Wang, Z.; Lv, Z.; Zhang, Z.; Wei, Z. Diode-Pumped High Energy and High Average Power All-Solid-State Picosecond Amplifier Systems. *Appl. Sci.* **2015**, *5*, 1590–1602. [CrossRef]

27. Roth, M.; Tseitlin, M.; Angert, N. Oxide crystals for electro-optic Q-switching of lasers. *Glass Phys. Chem.* **2005**, *31*, 86–95. [CrossRef]

28. Peng, Z.; Chen, M.; Yang, C.; Chang, L.; Li, G. A cavity-dumped and regenerative amplifier system for generating high-energy, high-repetition-rate picosecond pulses. *Jpn. J. Appl. Phys.* **2015**, *54*, 028001. [CrossRef]

29. Bai, Z.; Long, M.; Chen, L.; Chen, M.; Li, G. 145-watt high beam quality bidirectional voltage-supplied Q-switched Nd:YAG master oscillator power amplifier laser. *Opt. Eng.* **2013**, *52*, 024202. [CrossRef]

30. Wada, K.; Cho, Y. Improved expression for the time-bandwidth product of picosecond optical pulses from gain-switched semiconductor lasers. *Opt. Lett.* **1994**, *19*, 1633–1635. [CrossRef] [PubMed]

31. Bai, Z.; Bai, Z.; Yang, C.; Chen, L.; Chen, M.; Li, G. High pulse energy, high repetition picosecond chirped-multi-pulse regenerative amplifier laser. *Opt. Laser Technol.* **2013**, *46*, 25–28. [CrossRef]

*applied*
*sciences*

MDPI

*Article*

# Short-Pulse-Width Repetitively Q-Switched ~2.7-μm Er:Y$_2$O$_3$ Ceramic Laser

**Xiaojing Ren [1], Yong Wang [2], Jian Zhang [2], Dingyuan Tang [2] and Deyuan Shen [1,***

[1]   Department of Optical Science and Engineering, Fudan University, Shanghai 200433, China;
      xiaojingren099@126.com
[2]   Jiangsu Key Laboratory of Advanced Laser Materials and Devices, School of Physics and Electronic
      Engineering, Jiangsu Normal University, Xuzhou 221116, China; wangyong@jsnu.edu.cn (Y.W.);
      jzhang@jsnu.edu.cn (J.Z.); edytang@ntu.edu.cn (D.T.)
*     Correspondence: shendy@fudan.edu.cn; Tel.: +86-21-6564-2159

Received: 16 October 2017; Accepted: 17 November 2017; Published: 22 November 2017

**Abstract:** A short-pulse-width repetitively Q-switched 2.7-μm Er:Y$_2$O$_3$ ceramic laser is demonstrated using a specially designed mechanical switch, a metal plate carved with slits of both slit-width and duty-cycle optimized. With a 20% transmission output coupler, stable pulse trains with durations (full-width at half-maximum, FWHM) of 27–38 ns were generated with a repetition rate within the range of 0.26–4 kHz. The peak power at a 0.26 kHz repetition rate was ~3 kW.

**Keywords:** laser materials; mid-infrared lasers; rare-earth solid-state lasers

## 1. Introduction

Laser radiation at ~2.7-μm is important for practical applications and scientific research. Being regions of water absorption and molecular fingerprints, laser sources at ~2.7-μm are useful for biomedical therapy [1,2] and atmospheric sensing [3]. In addition, these lasers are utilized for generating 3–5-μm laser emission [4,5], which corresponds to an atmosphere transparent window. Lasers at ~2.7-μm also facilitate studies of laser materials that is suitable to generate deep-infrared lasing [6,7]. Both the applications and research require versatile ~2.7-μm laser sources with short pulse duration and high repetition rate.

A simple approach to generate ~2.7-μm pulses is Q-switching ion-based (as Er$^{3+}$, Ho$^{3+}$, Dy$^{3+}$ and Cr$^{2+}$) lasers. Er-based lasers operating on the transition between $^4I_{11/2}$ and $^4I_{13/2}$ energy levels are most often utilized because of the mature pump sources of flashlamps and ~976-nm laser diodes (LDs). With 50 atom % Er:YAG (Y$_3$Al$_5$O$_{12}$), 30 atom % Er:YSGG (Y$_3$Sc$_2$Ga$_3$O$_{12}$), and 15 atom % Er:YLF (LiYF$_4$) laser materials, ~2.7-μm pulses at durations of tens of nanoseconds have been obtained. However, the repetition rates of these lasers are generally limited to several Hertz due to the severely thermal effects generated during laser operations [8–11]. Laser pulses at ~2.7-μm with repetition rates at the kilohertz scale have been generated from Er:ZBLAN (ZrF$_4$-BaF$_2$-LaF$_3$-AlF$_3$-NaF) fiber lasers, which are more thermally advanced [12,13]. However, the pulse durations are hundreds of nanoseconds because fibers have limited energy storage capability. So far, short pulses at ~2.7-μm with high repetition rates are rare.

The use of Er-based sesquioxides is promising in terms of obtaining ~2.7-μm laser pulses with short pulse durations and high repetition rates. As ~2.7-μm laser oscillation can be realized from Er-based sesquioxides with low doping concentrations (lower than ~7 atom %) [14,15], the thermal effects generated during laser operation are greatly alleviated. In addition, sesquioxides have high thermal conductivity, which decreases slightly with increasing doping concentration [16]. Furthermore, Er-based sesquioxides have long ~2.7-μm fluorescence lifetimes (e.g., an order of magnitude longer than that of Er:YAG), making them beneficial for energy storage [17]. Unfortunately, sesquioxides have extremely high melting points (>2400 °C), imposing serious challenges for traditional single-crystal-growth approach. In this aspect, polycrystalline ceramics are superior to single crystals since they can be sintered at much lower temperatures. Moreover, transparent ceramics have advantages over crystals in terms of their rapid fabrication in large scale and composite structures, flexible doping concentrations, and good thermo-mechanical properties [18]. Recently, sesquioxide ceramics have been successfully fabricated and mainly explored to realize continuous-wave (CW) laser operation at ~2.7-µm. A passively Q-switched ~2.7-µm Er:Y$_2$O$_3$ ceramic laser with a pulse duration of 4.47 µs and a pulse repetition rate of 12.6 kHz was realized, aiming to demonstrate the broadband availability of black-phosphorus [19]. For actively Q-switched operation, an acousto-optically Q-switched Er:Y$_2$O$_3$ ceramic laser was demonstrated, generating pulses with durations of 41–190 ns in the range of 0.3–10 kHz [20]. Due to the scarcity of ~2.7-µm acousto-optic and electric-optic Q-switches, mechanical Q-switching is commonly used in a wavelength range of ~2.7-µm [21,22], which does not require high voltages or drive power and has avoidable insert losses.

Here we report a short-pulse-width repetitively Q-switched Er:Y$_2$O$_3$ ceramic laser at ~2.7-µm using a specially designed mechanical switch, a metal plate carved with slits of both slit-width and duty-cycle optimized. The laser performances with output couplers (OCs) of 5%, 8% and 20% transmissions are compared in both CW and Q-switched operation modes. In the CW operation mode, the 8% transmission OC yields an output power of over 1.8 W. In the Q-switched operation mode, the 20% transmission OC yields pulse trains with durations (FWHM) of 27–38 ns and energies of 80.8–27.5 µJ with repetition rates in the range of 0.26–4 kHz. The corresponding peak power at the 0.26 kHz repetition rate is ~3 kW. To the best of our knowledge, the laser offers the shortest pulse durations among ~2.7-µm Q-switched lasers with pulse repetition frequency (PRF) above several Hertz.

## 2. Experimental Details

Figure 1 shows the schematic layout of the pulsed Er:Y$_2$O$_3$ ceramic laser. A pig-tailed ~976-nm LD with a fiber core diameter of 105-µm and a numerical aperture (NA) of less than 0.15 was used to pump the Er:Y$_2$O$_3$ ceramic. The pump light was focused into the Er:Y$_2$O$_3$ sample with a ~420-µm spot diameter through a 25-mm-focal-length lens F1 and a 100-mm-focal-length lens F2. The confocal parameter was estimated to be ~26 mm. The physical length of the plane–plane cavity was 28 mm. The input coupler (IC) was high-transmission coated at ~976 nm (T > 98%) and high-reflectivity coated at ~2.7-µm (R > 99.8%). Three flat mirrors with transmissions of 5%, 8%, and 20% at ~2.7-µm and high transmissions at ~976 nm were employed as OCs. The Er:Y$_2$O$_3$ ceramic (developed at Jiangsu Normal university) was synthesized by the solid-state reaction method and vacuum sintering followed by hot isostatic pressing [14]. The ceramic sample had a dimension of 2 × 3 × 12 mm and an Er-ion concentration of 7 atom % and was uncoated. It was mounted in a copper block that was cooled by water at ~13 °C to allow for effective heat removal. A dichroic mirror (DM) coated with high reflectivity at the laser wavelength and high transmission at the pump wavelength was placed between the OC and the detector to filter out the unabsorbed pump power.

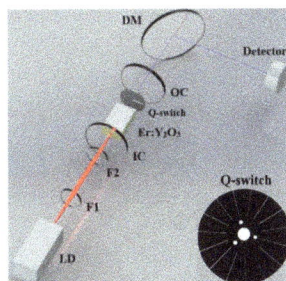

**Figure 1.** Schematic setup of the Q-switched Er:Y$_2$O$_3$ ceramic laser. LD: laser diode; IC: input coupler; OC: output coupler; DM: dichroic mirror.

The mechanical Q-switch was a 1-mm-thick rotating metal plate carved with sagittal rectangular slits (see the inset of Figure 1), of which the rotating speed and slit number were variable from 1 to 134 rounds per second and 1 to 30, respectively. It was placed close to the ceramic sample, where the diameter of the laser mode was calculated to be ~470-μm. The modulation frequency was adjusted by changing the slit number or (and) the rotating speed. The slit width was optimized to be 0.6 mm. The duty ratio of the Q-switch was continually tunable by moving the plate along its radial direction. Using such a mechanical Q-switch considerably increased the time for the inversion population to accumulate.

## 3. Results

A comparison of CW laser performances with 5%, 8% and 20% transmissions OCs of the Er:$Y_2O_3$ medium was first made. Figure 2 shows dependences of output powers on the absorbed pump power. The absorbed pump power was calculated by multiplying the incident pump power by the absorption efficiency, which was measured under non-lasing condition (without existence of the laser resonator) to be ~90%. The threshold pump powers for the 5%, 8% and 20% transmission OCs were ~1.1 W, ~1.4 W, and ~3 W, respectively. At pump powers of less than 14 W, the output powers increased linearly with the pump power for all three OCs, and the 8% OC yielded the highest slope efficiency of 10.2% with respect to absorbed pump power. When the pump power was above 14 W, the laser became slightly less efficient due to the increased thermally induced losses. Nevertheless, an output power of over 1.8 W was obtained at 19 W of pump power with the 8% transmission OC. According to the laser properties with different output couplers, the distributed cavity losses (excluding the output coupling loss) were calculated to be ~1.67% cm$^{-1}$ using the well-known Findlay–Clay method, comprising the thermally induced losses and the losses of the cavity mirrors and ceramic sample (~1.2% cm$^{-1}$ estimated though the in-line transmission spectrum of a 1.3-mm-long Er:$Y_2O_3$ sample). The attainable output power should be readily increased by improving the cooling system and reducing passive losses of the ceramic sample.

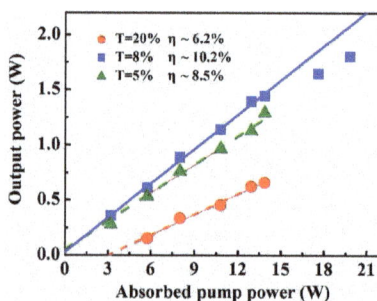

**Figure 2.** Continuous-wave (CW) output power vs. absorbed pump power with different output couplers.

The Q-switched operation performances were studied with the same output couplers. An HgCdTe detector (PVM-10.6, VigoSystem S. A) with a rise time of 1.5 ns and an oscilloscope (DSO104A, Keysight) with a bandwidth of 1 GHz were utilized to measure the pulse characters. The threshold for Q-switched operation was the same as that for the CW operation. Not too far above the threshold, stable pulse trains were observed for all three OCs. The pulse characters were then optimized by changing the distance between the center of the mechanical switch and the optical axes of the cavity, and the optimal distance was found to be ~50 mm. The pulse duration and pulse energy as functions of the pulse repetition frequency for all three OCs are shown in Figures 3 and 4. With the 5% and 8% transmission OCs, single pulses with 40 ns duration and 30.8 μJ energy, and 37 ns duration and 46.1 μJ energy were obtained at a 0.26 kHz pulse repetition frequency (corresponding to the rotating speed of

130 rounds per seconds and a slit number of 2) under a pump power of 3.2 W. With the increase in repetition frequency, the pulse duration increased while the pulse energy decreased. Nevertheless, pulse durations of less than 50 ns were obtained in the range of 0.26–4 kHz with both OCs. When further increasing the pump power, multi-pulse operation occurred. With the 20% transmission OC, shorter pulse duration and larger pulse energy were achieved. This is because a higher output coupling can delay the occurrence of multi-pulse operation in mechanically Q-switched lasers. In this case, multi-pulse operation occurred when the pump power was above 10 W. Under 10 W of pump power, the pulse duration increased from 27 to 38 ns, whereas the pulse energy decreased from 80.8 to 27.5 µJ with increasing the repetition frequency from 0.26 to 4 kHz, corresponding to ~3 kW peak power at 0.26 kHz repetition frequency. The optical–optical conversion efficiencies in the Q-switched mode could be effectively improved using a larger plate carved with more slits but the same turn-off area and/or a motor with higher rotating speed to make the laser operate at a higher repetition rate. Improvement in the Q-switched efficiencies is also achievable through optimizing the pump wavelength to avoid the excited state absorption of the upper laser level during the pumping process. The short pulse durations achieved in our experiment are a consequence of the good energy storage of Er:Y$_2$O$_3$, in which losses of excited ions on the upper laser level caused by multiphonon relaxation and spontaneous emission are small.

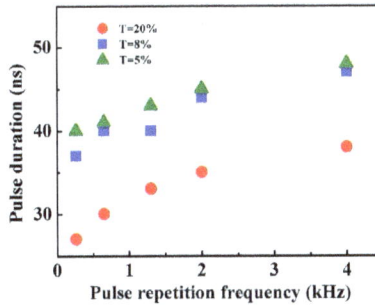

**Figure 3.** Pulse duration vs. pulse repetition frequency with different output couplers.

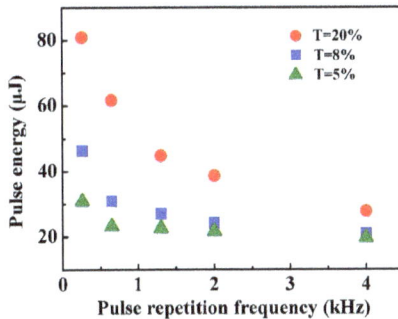

**Figure 4.** Pulse energy vs. pulse repetition frequency with different output couplers.

Figures 5 and 6 show the typical single pulse profile and pulse train recorded at 0.26 kHz repetition rate with the 20% transmission OC under 10 W of pump power. The pulse profile shows a fairly symmetric shape and the time spacing between two adjacent pulses remains nearly fixed for the given repetition rate with no noticeable timing jitter. The amplitude fluctuations were estimated to be less than 10%.

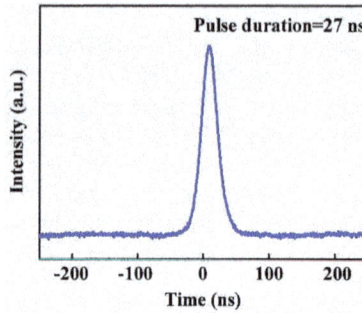

**Figure 5.** Typical pulse profile recorded at 0.26 kHz pulse repetition frequency (PRF) under 10 W of absorbed pump power.

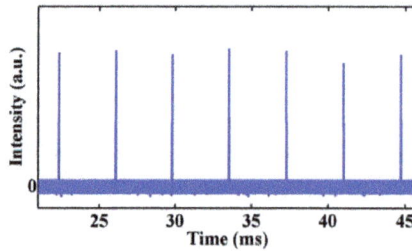

**Figure 6.** Pulse train recorded at 0.26 kHz PRF under 10 W of absorbed pump power.

The optical spectrum of the Q-switched Er:$Y_2O_3$ ceramic laser with the 20% transmission OC was found to vary slightly with the repetition rate or pump level. Figure 7 shows the typical emission wavelength measured by a Fourier transform spectrum analyzer with a resolution of 7.5 GHz (OSA205, Thorlabs, Newton, NJ, USA). It was centered at 2716.5 nm with a linewidth of 0.12 nm.

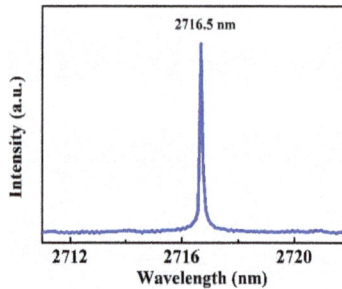

**Figure 7.** Spectrum of the Q-switched Er:$Y_2O_3$ laser.

## 4. Conclusions

In summary, we demonstrated a short-pulse-width repetitively Q-switched Er:$Y_2O_3$ ceramic laser at ~2.7-µm using a specially designed mechanical switch. Stable pulse trains with 27–38 ns durations and energies of 80.8–27.5 µJ were obtained with a pulse repetition frequency within a range of 0.26 to 4 kHz. A peak power of ~3 kW was realized at 0.26 kHz. The results presented here reveal the potential of Er-doped sesquioxide ceramics in generating short pulses at ~2.7-µm. Improvements in terms of pulse energy should be possible by further optimizing the transmission of the output coupler.

**Acknowledgments:** This work was supported by the National Science Foundation of China (NSFC) (61177045, 11274144), NSAF (U1430111), and a project funded by the Priority Academic Development of Jiangsu Higher Education Institutions (PAPD).

**Author Contributions:** Jian Zhang, Dingyuan Tang and Deyuan Shen conceived and designed the experiments; Xiaojing Ren and Yong Wang performed the experiments, analyzed the data, and wrote the paper.

**Conflicts of Interest:** The authors declare no conflict of interest.

## References

1. Guidotti, R.; Merigo, E.; Fornaini, C.; Rocca, J.P.; Medioni, E.; Vescovi, P. Er:YAG 2,940-nm laser fiber in endodontic treatment: A help in removing smear layer. *Lasers Med. Sci.* **2014**, *29*, 69–75. [CrossRef] [PubMed]
2. Zhao, Y.; Yin, Y.; Tao, L.; Nie, P.; Tang, Y.; Zhu, M. Er:YAG laser versus scaling and root planing as alternative or adjuvant for chronic periodontitis treatment: A systematic review. *J. Clin. Periodontol.* **2014**, *41*, 1069–1079. [CrossRef] [PubMed]
3. Ayerden, N.P.; Mandon, J.; Ghaderi, M.; Harren, F.J.; Wolffenbuttel, R.F. A photonic microsystem for hydrocarbon gas analysis by mid-infrared absorption spectroscopy. In Proceedings of the 2017 IEEE 30th International Conference on Micro Electro Mechanical Systems (MEMS), Las Vegas, NV, USA, 22–26 January 2017; pp. 1052–1055.
4. Gauthier, J.C.; Fortin, V.; Carree, J.Y.; Poulain, S.; Poulain, M.; Vallee, R.; Bernier, M. Mid-IR supercontinuum from 2.4 to 5.4 µm in a low-loss fluoroindate fiber. *Opt. Lett.* **2016**, *41*, 1756–1759. [CrossRef] [PubMed]
5. Robichaud, L.R.; Fortin, V.; Gauthier, J.C.; Chatigny, S.; Couillard, J.F.; Delarosbil, J.L.; Vallee, R.; Bernier, M. Compact 3–8 µm supercontinuum generation in a low-loss $As_2Se_3$ step-index fiber. *Opt. Lett.* **2016**, *41*, 4605–4608. [CrossRef] [PubMed]
6. Vodopyanov, K.L.; Makasyuk, I.; Schunemann, P.G. Grating tunable 4–14 µm GaAs optical parametric oscillator pumped at 3 µm. *Opt. Express* **2014**, *22*, 4131–4136. [CrossRef] [PubMed]
7. Frolov, M.P.; Korostelin, Y.V.; Kozlovsky, V.I.; Podmar'kov, Y.P.; Savinova, S.A.; Skasyrsky, Y.K. 3 J pulsed Fe:ZnS laser tunable from 3.44 to 4.19 µm. *Laser Phys. Lett.* **2015**, *12*, 055001. [CrossRef]
8. Zajac, A.; Skorczakowski, M.; Swiderski, J.; Nyga, P. Electrooptically Q-switched mid-infrared Er:YAG laser for medical applications. *Opt. Express* **2004**, *12*, 5125–5130. [CrossRef] [PubMed]
9. Högele, A.; Hörbe, G.; Lubatschowski, H.; Welling, H.; Ertmer, W. 2.70 µm CrEr:YSGG laser with high output energy and FTIR-Q-switch. *Opt. Commun.* **1996**, *125*, 90–94. [CrossRef]
10. Maak, P.; Jakab, L.; Richter, P.; Eichler, H.J.; Liu, B. Efficient acousto-optic Q switching of Er:YSGG lasers at 2.79-µm wavelength. *Appl. Opt.* **2000**, *39*, 3053–3059. [CrossRef] [PubMed]
11. Wyss, C.H.R.; Luthy, W.; Weber, H.P. Modulation and single-spike switching of a diode-pumped $Er^{3+}$:LiYF$_4$ laser at 2.8 µm. *IEEE J. Quantum Electron.* **1998**, *34*, 1041–1045. [CrossRef]
12. Qin, Z.; Xie, G.; Zhang, H.; Zhao, C.; Yuan, P.; Wen, S.; Qian, L. Black phosphorus as saturable absorber for the Q-switched Er:ZBLAN fiber laser at 2.8 µm. *Opt. Express* **2015**, *23*, 24714–24718. [CrossRef] [PubMed]
13. Shen, Y.; Wang, Y.; Luan, K.; Huang, K.; Tao, M.; Chen, H.; Yi, A.; Feng, G.; Si, J. Watt-level passively Q-switched heavily $Er^{3+}$-doped ZBLAN fiber laser with a semiconductor saturable absorber mirror. *Sci. Rep.* **2016**, *6*, 26659. [CrossRef] [PubMed]
14. Qiao, X.B.; Huang, H.T.; Yang, H.; Zhang, L.; Wang, L.; Shen, D.Y.; Zhang, J.; Tang, D.Y. Fabrication, optical properties and LD-pumped 2.7 µm laser performance of low $Er^{3+}$ concentration doped $Lu_2O_3$ transparent ceramics. *J. Alloys Compd.* **2015**, *640*, 51–55. [CrossRef]
15. Sanamyan, T. Diode pumped cascade Er:Y$_2$O$_3$ laser. *Laser Phys. Lett.* **2015**, *12*, 125804. [CrossRef]
16. Krankel, C. Rare-Earth-Doped Sesquioxides for Diode-Pumped High-Power Lasers in the 1-, 2-, and 3-µm Spectral Range. *IEEE J. Sel. Top. Quantum Electron.* **2015**, *21*, 1602013. [CrossRef]
17. Sanamyan, T.; Simmons, J.; Dubinskii, M. $Er^{3+}$-doped Y$_2$O$_3$ ceramic laser at ~2.7 µm with direct diode pumping of the upper laser level. *Laser Phys. Lett.* **2010**, *7*, 206–209. [CrossRef]
18. Ikesue, A.; Aung, Y.L. Ceramic laser materials. *Nat. Photonics* **2008**, *2*, 721–727. [CrossRef]
19. Kong, L.C.; Qin, Z.P.; Xie, G.Q.; Guo, Z.N.; Zhang, H.; Yuan, P.; Qian, L.J. Black phosphorus as broadband saturable absorber for pulsed lasers from 1 µm to 2.7 µm wavelength. *Laser Phys. Lett.* **2016**, *13*, 045801. [CrossRef]

20. Ren, X.J.; Wang, Y.; Fan, X.L.; Zhang, J.; Tang, D.Y.; Shen, D.Y. High-peak-power acousto-optically Q-switched Er:$Y_2O_3$ ceramic laser at ~2.7 μm. *IEEE Photonics J.* **2017**, *9*, 1–6. [CrossRef]

21. Skórczakowski, M.; Pichola, W.; Šwiderski, J.; Nyga, P.; Galecki, L.; Maciejewska, M.; Kasprzak, J. 30 mJ, TEM$_{00}$, high repetition rate, mechanically Q-switched Er:YAG laser operating at 2940 nm. *Opto-Electron. Rev.* **2011**, *19*, 206–210. [CrossRef]

22. Murphy, F.J.; Arbabzadah, E.A.; Bak, A.O.; Amrania, H.; Damzen, M.J.; Phillips, C.C. Optical chopper Q-switching for flashlamp-pumped Er,Cr:YSGG lasers. *Laser Phys. Lett.* **2015**, *12*, 045802. [CrossRef]

*applied*
*sciences*

MDPI

Article

# A High-Power Continuous-Wave Mid-Infrared Optical Parametric Oscillator Module

Yichen Liu [1], Xukai Xie [1], Jian Ning [1,2], Xinjie Lv [1,2,*], Gang Zhao [1,2], Zhenda Xie [1,3,*] and Shining Zhu [1]

[1]   National Laboratory of Solid State Microstructures, Nanjing University, Nanjing 210093, China;
      yourslyc@163.com (Y.L.); xiexukai3@163.com (X.X.); ningjian1991@163.com (J.N.);
      zhaogang@nju.edu.cn (G.Z.); zhusn@nju.edu.cn (S.Z.)
[2]   College of Engineering and Applied Sciences, Nanjing University, Nanjing 210093, China
[3]   School of Electronic Science and Engineering, Nanjing University, Nanjing 210093, China
*    Correspondence: lvxinjie@nju.edu.cn (X.L.); xiezhenda@nju.edu.cn (Z.X.);
     Tel.: +86-25-8359-4660 (X.L.); +86-25-8362-1225 (Z.X.)

Received: 31 October 2017; Accepted: 12 December 2017; Published: 21 December 2017

**Abstract:** We demonstrate here a compact optical parametric oscillator module for mid-infrared generation via nonlinear frequency conversion. This module weighs only 2.5 kg and fits within a small volume of $220 \times 60 \times 55$ mm$^3$. The module can be easily aligned to various pump laser sources, and here we use a 50 W ytterbium (Yb)-doped fiber laser as an example. With a two-channel MgO-doped periodically poled lithium niobate crystal (MgO:PPLN), our module covers a tuning range of 2416.17–2932.25 nm and 3142.18–3452.15 nm. The highest output power exceeds 10.4 W at 2.7 μm, corresponding to a conversion efficiency of 24%. The measured power stability is 2.13% Root Meat Square (RMS) for a 10 h duration under outdoor conditions.

**Keywords:** mid-infrared laser; optical parametric oscillator; laser module

## 1. Introduction

Coherent mid-infrared (MIR) radiation is important for applications in many fields, such as gas sensing, ranging, spectroscopy, biochemistry, atmospheric science, and security [1–3]. Direct MIR generation can be achieved using quantum cascade lasers, but the output power, and thus the sensing and interaction distance, is limited [4,5]. On the other hand, high power MIR radiation can be generated from nonlinear optical frequency conversion in the form of optical parametric oscillators (OPOs) [6–9], and an output power up to 10 W has been reported [10]. However, most OPOs are built on optical tables with bulky optics and thus are not compatible for portable applications, such as moving vehicles based on the air, land or water [11–13].

In this article, we demonstrate an OPO module, which is embedded in a monolithic metal frame, with dimensions and a weight of $220 \times 60 \times 55$ mm$^3$ and 2.5 kg, respectively. Such a module is compatible, and can be easily aligned, with any continuous-wave laser centered around 1064 nm for the MIR generation, and an ytterbium-doped fiber laser pump was used for our test. The MIR tuning ranges were 2416.17–2932.25 nm and 3142.18–3452.15 nm, using two channels of a periodically poled lithium niobate (PPLN) crystal with poling periods of 31.59 and 30.49 μm channels, respectively. At a maximum pump power of 50 W, the output power exceeded 10.4 and 9.1 W at 2.7 and 3.3 μm, respectively. The 10 h power stability was measured to be 2.13% root meat square (RMS) in a room without any special temperature stabilization.

## 2. Module Design

The frame of the MIR module was made from a single block of 7075 aluminum alloy for its high strength-to-weight ratio. As a result, we managed to construct the whole module with a weight of less than 2.5 kg. The configuration and schematic of our module is shown in Figure 1, and it was designed to be quickly aligned with any continuous-wave (CW) pump laser at around 1064 nm. The pump path into the OPO cavity was aligned to two irises in our module, and two high-reflection mirrors R1 and R2 can be used to steer the pump beam for the input alignment. Once the pump light matches with the irises, the module is ready for MIR generation. Inside the module is a symmetric ring OPO cavity. Cavity mirrors M1–M4 are all high-reflection-coated (R > 99.5%) at 1.35–2 μm (signal wavelength) and high-transmission-coated at 1064 nm (pump wavelength, T > 95%) and 2.3–5 μm (idler wavelength, T > 90%). M1 and M2 on both ends of the short arm are plane mirrors, while M3 and M4 on the long arm are concave mirrors with a curvature radius of 100 mm. The nonlinear medium is a periodically poled lithium niobate crystal (MgO:PPLN) doped with 5 mol % MgO with dimensions of $50 \times 10 \times 1$ mm$^3$ and two channels with poling periods of 30.49 and 31.59 μm, for the dual band MIR generation around 2.7 and 3.3 μm, respectively. Both ends of the crystal are antireflection-coated at pump, signal, and idler wavelengths. The crystal temperature is controlled using a homemade oven, where the crystal mount is temperature-controlled with an accuracy of ±0.1 °C in a range between 20 and 170 °C. It is suspended from the oven enclosure by three ceramic tubes, which form a 2 mm air gap to reduce the thermal conductivity and further stabilize the temperature of MgO:PPLN. The surface of the crystal mount is finely ground for good thermal contact with the MgO:PPLN. A homemade miniaturized linear translation stage and electric actuator were built into the temperature-controlled oven for channel switching for the MgO:PPLN, and together they weigh only 10 g.

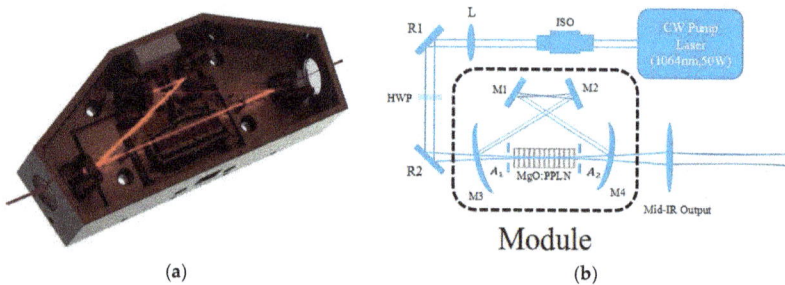

(a)

(b)

**Figure 1.** The configuration of the module (**a**) and the optical schematic of the experiment (**b**). HWP: half wavelength plate; ISO: isolator; MgO:PPLN: MgO-doped periodically poled lithium niobate crystal; CW: continuous-wave.

## 3. Experiment and Results

In this experiment, we pumped our OPO module with a CW, linearly polarized, single-frequency Yb-doped fiber laser (IPG Photonics). The laser can deliver a maximum power of 50 W at 1064 nm with a beam diameter of 2 mm. Its spectral linewidth is only 70 kHz, which is one of the keys to achieving a narrowband MIR output. A free space optical isolator was used to protect the pump laser from unwanted back scattering, followed by a half-wave plate (HWP) to recover the polarization of 45° to a vertical polarization, and matched the Z-axis of the MgO:PPLN. A focal lens L with a focal length of 150 mm beam was used to focus the pump light to the center of the MgO:PPLN, with a beam waist radius of 58 μm. Considering the different diffraction properties at pump and signal wavelengths, we made the radius of signal light slightly larger than that of the pump light to optimize the overlap mode between them.

We measured the temperature tuning from the OPO module with both channels by varying the MgO:PPLN temperature from 20 to 170 °C. The MIR wavelength can be tuned from 2416.17 to 2932.25 nm or from 3142.18 to 3452.15 nm in 31.59-μm- and 30.49-μm-poled channels, as shown by the black and red curves in Figure 2, respectively.

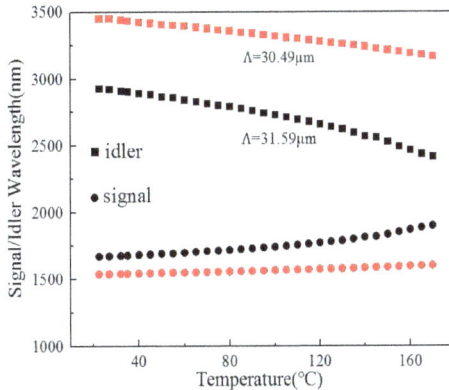

**Figure 2.** The output wavelength of the module as a function of the crystal temperature.

There are two wavelengths of special interest in the tuning range: 2.7 and 3.3 μm, which correspond to the absorption peaks of H–O and C–H bonds, respectively. The strong molecular rovibrational transitions that can be accessed with mid-IR laser sources allowed trace-gas sensing down to the parts-per-quadrillion level. We measured the output power from the MIR module at these two wavelengths. Figure 3 shows the output power as functions of the pump power when the MgO:PPLN temperature is set at 105.6 °C. The MIR wavelength was 2.7 μm for the 31.59-μm-poled channel, corresponding to a signal wavelength of 1.75 μm. For the 30.49 μm channel, the MIR and signal wavelength were 3.3 and 1.56 μm, respectively. The thresholds of the two above processes were 5.9 and 7.4 W, respectively. The output power for both wavelengths rose steadily as the pump power increased. The MIR output was limited by the maximum pump power of 50 W and exceeded 10.4 and 9.1 W at 2.7 and 3.3 μm, respectively. The maximum conversion efficiencies were 24% for 2.7 μm and 21% for 3.3 μm, respectively. The conversion efficiencies of the idler output were not linear because of the saturation of the pump depletion.

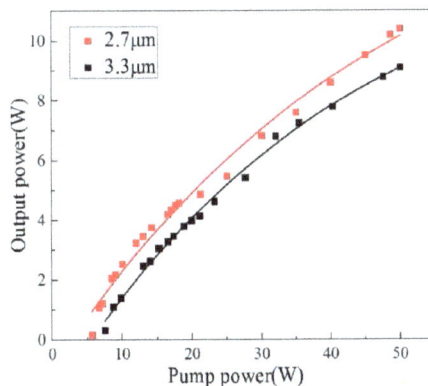

**Figure 3.** Output power of the module at 2.7 and 3.3 μm as a function of incident pump power.

The power stability of the OPO module was measured. The test lasted 10 h in a room without special temperature stabilization. The pump power was maintained at 35 W throughout the entire testing process. The MIR power was about 6.2 W at this pump power. As shown in Figure 4, the RMS of the MIR output power in 10 h was 2.13%. We measured the spectra characteristic using an optical spectra analyzer (YOKOGAWA AQ6375, YOKOGAWA, Tokyo, Japan), where the measured linewidth was limited by the instruction spectral resolution of 6 GHz. However, we could infer that our OPO was oscillating in a single longitudinal mode.

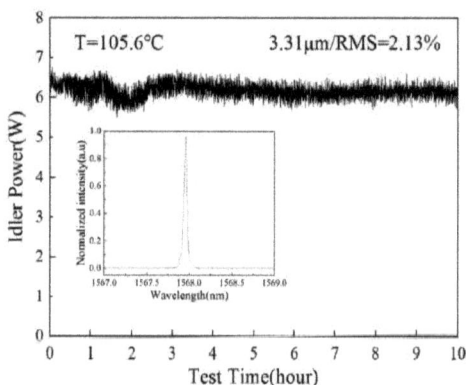

**Figure 4.** The long time power stability of 3.31 μm. Inset: the spectrum of the signal.

## 4. Conclusions

In summary, we have designed and fabricated an OPO module. The module is compact and confined, with dimensions and a weight of $220 \times 60 \times 55$ mm$^3$ and 2.5 kg. The experiment results show that the module had a wide tuning range, a high power, and a stable output when a commercial Yb fiber laser was used as the pump laser. The MIR tuning ranges were 2416.17–2932.25 nm and 3142.18–3452.15 nm when two channels of a periodically poled lithium niobate (PPLN) crystal, with poling periods of 31.59 and 30.49 μm channels, respectively, were used. At the maximum pump power (50 W), the highest output power of the module exceeded 10.4 W at 2.7 μm and 9.1 W at 3.3 μm. To our knowledge, this is the best performing CW-OPO based on PPLN. The 10 h power stability at 3.31 μm was measured to be 2.13% RMS under outdoor environment. We believe that such a compact, tunable, stable, and high-power OPO module can be used as the core part of a number of mid-infrared laser sources. Its promoting effect in the relevant industries is foreseeable.

**Acknowledgments:** This work was supported by the National Key Research and Development Program of China (No. 2017YFB0405200), the National Young 1000 Talent Plan, National Natural Science Foundation of China (No. 91321312, No. 11621091, No. 11674169), the Ministry of Science and Technology of the People's Republic of China (No. 2017YFA0303700), the International Science and Technology Cooperation Program of China (ISTCP) (No. 2014DFT50230), the Key Research Program of Jiangsu Province (No. BE2015003-2), and Special Funds for Fundamental Scientific Research Business Fees in Central Universities.

**Author Contributions:** Zhenda Xie and Xinjie Lv conceived and designed the experiments. Yichen Liu, Jian Ning, Gang Zhao and Xukai Xie performed the experiments. Yichen Liu and Zhenda Xie wrote the paper, and Shining Zhu supervise the whole work.

**Conflicts of Interest:** The authors declare no conflict of interest.

## References

1. Asobe, M.; Tadanaga, O.; Umeki, T.; Yanagawa, T.; Magari, K.; Ishii, H. Engineered Quasi-Phase Matching Device for Unequally Spaced Multiple Wavelength Generation and its Application to Midinfrared Gas Sensing. *IEEE J. Quantum Electron.* **2010**, *46*, 447–453. [CrossRef]

2. Arslanov, D.D.; Spunei, M.; Mandon, J.; Cristescu, S.M.; Persijn, S.T.; Harren, F.J.M. Continuous-wave optical parametric oscillator based infrared spectroscopy for sensitive molecular gas sensing. *Laser Photonics Rev.* **2013**, *7*, 188–206. [CrossRef]

3. Henderson, A.; Stafford, R.; Miller, J.H. Continuous wave optical parametric oscillators break new spectral ground. *Spectroscopy* **2005**, *20*, 16–18.

4. Borri, S.; de Cumis, M.S.; Insero, G.; Santambrogio, G.; Savchenkov, A.; Eliyahu, D.; Ilchenko, V.; Matsko, A.; Maleki, L.; De Natale, P. Whispering gallery mode stabilization of quantum cascade lasers for infrared sensing and spectroscopy. In Proceedings of the Conference on Laser Resonators, Microresonators, and Beam Control XIX, San Francisco, CA, USA, 30 January–2 February 2017.

5. Yao, Y.; Hoffman, A.J.; Gmachl, C.F. Mid-infrared quantum cascade lasers. *Nat. Photonics* **2012**, *6*, 432–439. [CrossRef]

6. Tsai, L.; Chen, Y.F.; Lin, S.; Lin, Y.; Huang, Y. Compact efficient passively Q-switched Nd: GdVO$_4$/PPLN/Cr$^{4+}$: YAG tunable intracavity optical parametric oscillator. *Opt. Express* **2005**, *13*, 9543–9547. [CrossRef] [PubMed]

7. Oshman, M.K.; Harris, S. Theory of optical parametric oscillation internal to the laser cavity. *IEEE J. Quantum Electron.* **1968**, *4*, 491–502. [CrossRef]

8. Colville, F.G.; Dunn, M.H.; Ebrahimzadeh, M. Continuous-wave, singly resonant, intracavity parametric oscillator. *Opt. Lett.* **1997**, *22*, 75–77. [CrossRef] [PubMed]

9. Myers, L.E.; Eckardt, R.C.; Fejer, M.M.; Byer, R.L.; Bosenberg, W.R.; Pierce, J.W. Quasi-phase-matched optical parametric oscillators in bulk periodically poled LiNbO3. *J. Opt. Soc. Am. B* **1995**, *12*. [CrossRef]

10. Chen, D.-W.; Rose, T.S. Low noise 10-W cw OPO generation near 3 μm with MgO doped PPLN. In Proceedings of the Conference on Lasers and Electro-Optics (CLEO), San Jose, CA, USA, 22–27 May 2005.

11. Bosenberg, W.R.; Drobshoff, A.; Alexander, J.I.; Myers, L.E.; Byer, R.L. 93% pump depletion, 3.5-W continuous-wave, singly resonant optical parametric oscillator. *Opt. Lett.* **1996**, *21*, 1336–1338. [CrossRef] [PubMed]

12. Vainio, M.; Peltola, J.; Persijn, S.; Harren, F.J.M.; Halonen, L. Singly resonant cw OPO with simple wavelength tuning. *Opt. Express* **2008**, *16*, 11141–11146. [CrossRef] [PubMed]

13. Ricciardi, I.; De Tommasi, E.; Maddaloni, P.; Mosca, S.; Rocco, A.; Zondy, J.-J.; De Rosa, M.; De Natale, P. A narrow-linewidth optical parametric oscillator for mid-infrared high-resolution spectroscopy. *Mol. Phys.* **2012**, *110*, 2103–2109. [CrossRef]

*applied*
*sciences*

MDPI

Article

# Design of 4 × 1 Power Beam Combiner Based on MultiCore Photonic Crystal Fiber

Dror Malka [1,*], Eyal Cohen [2] and Zeev Zalevsky [2]

[1]   Faculty of Engineering, Holon Institute of Technology (HIT), Holon 5810201, Israel
[2]   Faculty of Engineering, Bar Ilan University, Ramat-Gan 52900, Israel; eyalco28@yahoo.com (E.C.);
      zalevsz@biu.ac.il (Z.Z.)
*    Correspondence: drorm@hit.ac.il; Tel.: +972-350-266-48

Academic Editor: Federico Pirzio
Received: 14 June 2017; Accepted: 3 July 2017; Published: 5 July 2017

**Featured Application: The proposed device can be very useful to high power and parametric applications.**

**Abstract:** A novel concept of 4 × 1 power beam combiner based on multicore photonic crystal fiber is described. The light coupling obtained by integrating small air-holes in the multicore photonic crystal fiber (PCF) structure allows light coupling between coherent laser sources to the central core. The beam propagation method (BPM) and coupled mode theory were used for analyzing the proposed device. Simulation results show that four coherent fiber laser sources of 1 µm in a multicore PCF structure can be combined into one source after 2.6 mm light propagation, with a power efficiency of 99.6% and bandwidth of 220 nm. In addition, a higher 8 × 1 ratio combiner was demonstrated, based on the proposed device. Thus, the device can be very useful to combine beams.

**Keywords:** combiner; beam propagation method; photonic crystal fiber

## 1. Introduction

In 1961, the first laser based on neodymium ($Nd^{3+}$) doped fiber was demonstrated [1,2]. Since then, fiber laser has been found as an efficient source for various applications, such as laser material processing [3], medical diagnostics [4], high power [5], metrology [6], imaging [7], etc. The most commonly used laser-active ion dopants in fibers are erbium ($Er^{3+}$) and ytterbium ($Yb^{3+}$) dopants, with their emission wavelengths around 1.5–1.6 µm and 1–1.1 µm, respectively [8,9].

The benefits of using a fiber laser as a light source: strong stability against thermo-optic effects, ease of use and higher gain, can be obtained by long fiber, while still keeping a compact cavity structure. Another advantage of fiber lasers is the capability to achieve output power values as high as 10 kW [10], using a single fiber laser. However, problems such as modal instabilities, thermal damage, and nonlinear effects, limit the power levels of a single fiber laser [11]. In order to overcome these problems, several beam combining methods were developed, such as the coherent beam combining [12], spectral beam combining [13], and incoherent beam combining [14].

Another solution is to use the combiner based on photonic crystal fiber (PCF). PCF is a versatile technology, based on a microstructured formation of low- and high-index materials [15]. Usually, the background material is pure silica (high-index), and the low-index areas are air holes along the fiber length.

PCF has unique characteristics [16,17] that do not exist in classical fibers such as high birefringence, larger single-mode areas, extremely low/high nonlinearity, and lower coupling length value between two closer cores. In recent years, research has demonstrated the potential of using PCF based coupler/splitter devices [17–19]. One of the recent improvements is the ability to obtain a smaller value of the coupling length compared to coupler/splitter based classical fibers.

Recently, the authors demonstrated a compact 4 × 1, 8 × 1 and 16 × 1 power combiner based on PCF [20]. The combining was achieved by replacing some air-holes areas with pure silica along the fiber length. However, this approach cannot be fabricated by fiber drawing methods [21–23]. In order to solve this problem, we propose a new approach that involves small air-holes in the PCF structure, which allows the control of light coupling between close cores without changing the refractive index structure. Thus, this technique can be used with the drawing and stack method for fabricating a combiner device based on multicore PCF. In this paper, we propose a new approach to obtain a multicore PCF that combined multiple laser sources to one source with high power level. The coupled mode theory and the beam propagation method [24,25] were used to investigate the performances of the 4 × 1 power combiner. In addition, a higher ratio of 8 × 1 combiner was designed using a cascade of two 4 × 1 power combiners and one 2 × 1 power combiner.

## 2. Principle of the Work

Figure 1 shows a schematic sketch of the multicore PCF 4 × 1 power combiner design based on the coupling between five cores (yellow color). The coefficients $\kappa_1$, $\kappa_2$ are the coupling coefficients between the cores: $\kappa_1$ denotes the coupling between core 3 and core 1, or between core 2 and core 1. $\kappa_2$ denotes the coupling between core 4 and core 1, or between core 5 and core 1. $d$ is the diameter of the air-holes (white color), $d$ is the diameter of the small air-holes (black color) and $\Lambda$ denotes the pitch—the distance between two air holes.

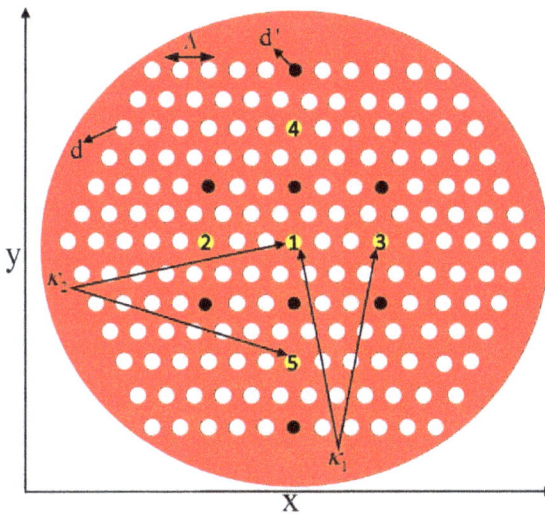

**Figure 1.** Schematic sketch of the 4 × 1 power combiner design at the *x–y* plane.

Classical coupled-mode equations can be used to analyze the light coupling between the five cores in the multicore PCF 4 × 1 power combiner. In our model, we integrated eight small air holes (black color), which leads to cancellation of other coupling that may occur between closer cores such as core 3–core 5, core 3–core 4, core 2–core 4, and core 2–core 5, and by assuming polarization independence, the solutions of the mode amplitudes can be given as

$$A_r = \alpha_n e^{(-jz(\beta+\varepsilon))} \tag{1}$$

where $A_r$ ($r = 1, 2, 3, 4, 5$) are the amplitudes of the fundamental mode in core $r$, $\beta$ is the propagation constant of the fundamental propagation mode, $z$ is the propagation distance, $\alpha_n$ ($n = 1, 2, 3, 4, 5$) is the amplitude constant, and $\varepsilon$ is an eigenvalue. In our design, there are five cores, which mean that five coupled-mode equations can be used to describe our design. However, the combiner is based on multicore PCF with symmetrical hexagonal structure; in other words, the mode amplitude is equal in core 2–core 3 ($A_2 = A_3$) and core 4–core 5 ($A_4 = A_5$). Therefore, the five coupled-mode equations can be reduced to three equations which are given by

$$\frac{d(\alpha_1 e^{(-jz(\beta+\varepsilon))})}{dz} + j\beta\alpha_1 e^{(-jz(\beta+\varepsilon))} = -j\left\{2\alpha_2 e^{(-jz(\beta+\varepsilon))}\kappa_1 + 2\alpha_4 e^{(-jz(\beta+\varepsilon))}\kappa_2\right\} \tag{2}$$

$$\frac{d(\alpha_2 e^{(-jz(\beta+\varepsilon))})}{dz} + j\beta\alpha_2 e^{(-jz(\beta+\varepsilon))} = -j\alpha_1 e^{(-jz(\beta+\varepsilon))}\kappa_1 \tag{3}$$

$$\frac{d(\alpha_4 e^{(-jz(\beta+\varepsilon))})}{dz} + j\beta\alpha_4 e^{(-jz(\beta+\varepsilon))} = -j\alpha_1 e^{(-jz(\beta+\varepsilon))}\kappa_2 \tag{4}$$

where the boundary conditions are given by

$$A_1(z = 0) = 0, A_2(z = 0) = A_4(z = 0) = 0.25 \tag{5}$$

The three coupled-mode Equations (2)–(4) can be simplified to three linear equations, which are given by

$$\varepsilon\alpha_1 = 2\alpha_2\kappa_1 + 2\alpha_4\kappa_2 \tag{6}$$

$$\varepsilon\alpha_2 = \alpha_1\kappa_1 \tag{7}$$

$$\varepsilon\alpha_4 = \alpha_1\kappa_2 \tag{8}$$

The matrix system can be described as follows:

$$\begin{bmatrix} \varepsilon & -2\kappa_1 & -2\kappa_2 \\ -\kappa_1 & \varepsilon & 0 \\ -\kappa_2 & 0 & \varepsilon \end{bmatrix} \begin{bmatrix} \alpha_1 \\ \alpha_2 \\ \alpha_4 \end{bmatrix} = \begin{bmatrix} 0 \\ 0 \\ 0 \end{bmatrix} \tag{9}$$

The eigenvalues and the eigenvectors can be found by solving the matrix system (Equation (9)). The field $E(z)$ can be represented by a linear combination of the eigenvectors. In a particular solution, where $\kappa_1 = \kappa_2$, a complete transfer of the energy from cores 2, 3, 4, and 5 to the central core 1, can be obtained. However, this condition depends on the geometrical parameter values of the multicore PCF structure ($z$, $d$, $d'$, and $\Lambda$). Therefore, optimization of the key parameters was done, in order to fulfill the necessary condition. In addition, this model can be duplicated, in order to design a higher ratio power such as 8 × 1 combiner.

## 3. Results: The Designs of 4 × 1 and 8 × 1 Power Combiners

Figures 2a–c and 3a–c shows the refractive index structure of the 4 × 1 power combiner and 8 × 1 power combiner, respectively. In these figures, the red color areas represent silica, and the purple color areas represent air.

The optimal values of the 4 × 1 power combiner multicore PCF structure are

$$z = 2.6 \text{ mm}, \ d = 0.8277 \ \mu\text{m}, \ d' = 0.2483 \ \mu\text{m}, \ \Lambda = 2.365 \ \mu\text{m}, \ \frac{d'}{\Lambda} = 0.105, \ \frac{d}{\Lambda} = 0.35$$

The 8 × 1 power combiner is based on a cascade which includes two units of 4 × 1 power combiners and one unit of 2 × 1 power combiner, as shown in Figure 3a–c.

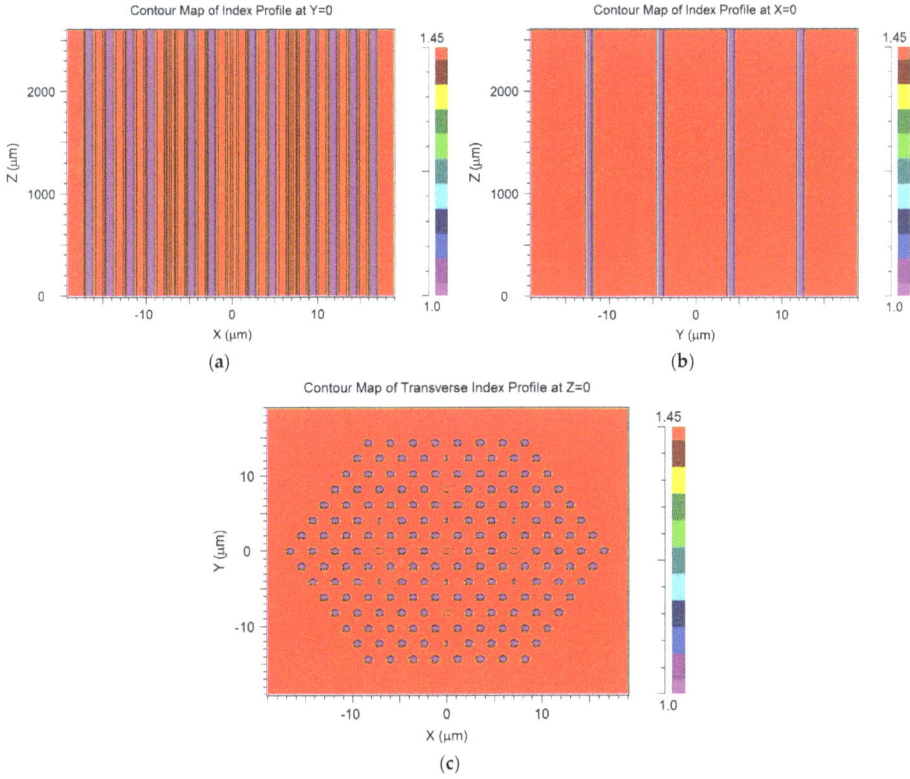

**Figure 2.** Refractive index profiles of the 4 × 1 power combiner: (**a**) *xz* plane at *y* = 0; (**b**) *yz* plane at *x* = 0; (**c**) *xy* plane at *z* = 0.

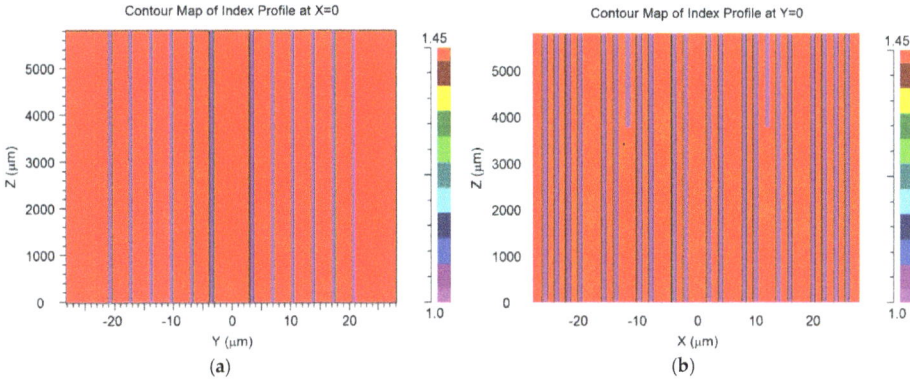

**Figure 3.** *Cont.*

Contour Map of Transverse Index Profile at Z=0

(c)

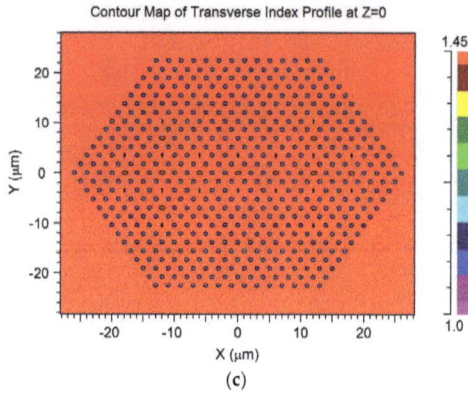

**Figure 3.** Refractive index profiles of the 8 × 1 power combiner: (**a**) *xz* plane at *y* = 0; (**b**) *yz* plane at *x* = 0; (**c**) *xy* plane at *z* = 0.

## 4. Simulation Results

The multicore PCF 4 × 1 power combiner structure was simulated using RSoft Photonics CAD software (5.1.5, RSoft, Ossining, NY, USA), based on BPM.

Figure 4a shows the transmission of four Gaussian sources at a 1 μm wavelength at *z* = 0, with a normalized power value of 0.25. Figure 4b shows the light coupling between the central core to the other four cores at *z* = 1 mm. Figure 4c shows that four Gaussian sources are combined to one source at *z* = 2.6 mm, with 99.6% of the total power.

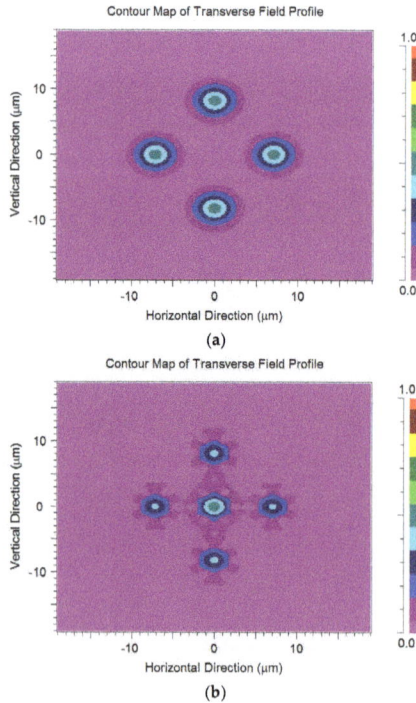

Contour Map of Transverse Field Profile

(a)

Contour Map of Transverse Field Profile

(b)

**Figure 4.** *Cont.*

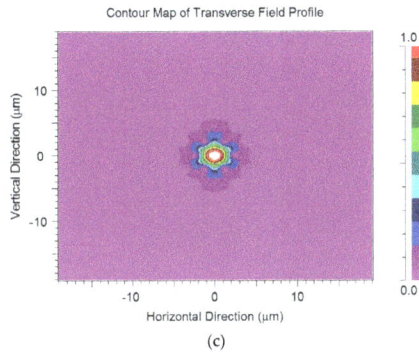

**Figure 4.** The 4 × 1 power combiner of the optical signals (λ = 1 μm) of three *xy* cross sections: (**a**) *z* = 0 mm, (**b**) *z* = 1 mm; (**c**) *z* = 2.6 mm.

The proposed device can also act as a two dimensional (2D) 2 × 1 power combiner, as shown in Figure 5a,b.

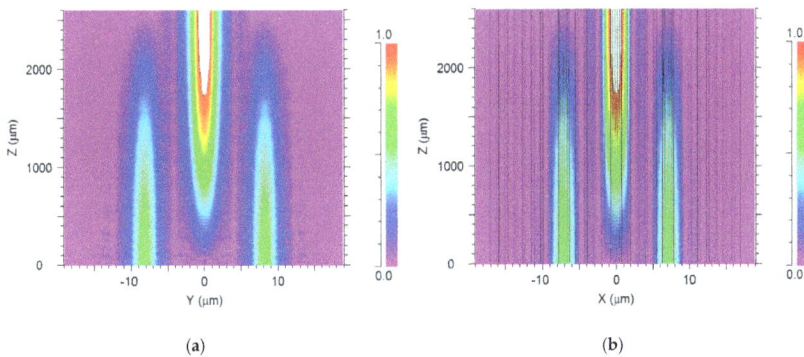

**Figure 5.** Two dimensional (2D) 2 × 1 power combiner of the optical signals (λ= 1 μm): (**a**) *yz* plane; (**b**) *xz* plane.

In addition, a MATLAB script code, combined with BPM simulations, was developed to examine the sensitivity of the proposed device to the wavelength variation of the laser sources around 1 μm. Figure 6 shows power attenuation around the central wavelength (1 μm).

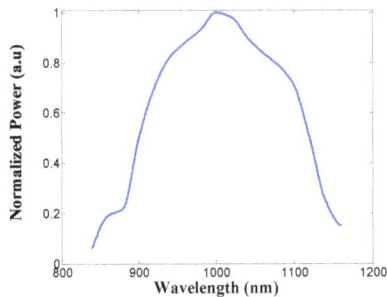

**Figure 6.** Normalized power in the central core at *z* = 2.6 mm as a function of wavelength.

It can be noticed from Figure 6 that the bandwidth (FWHM) of the 4 × 1 power combiner is about 220 nm in the 900–1120 nm range. Such a bandwidth implies that this combiner may be suitable for tunable lasers around a wavelength of 1000 nm, and can support broadband sources. This provides many benefits, due to the fact that within this range, one can use optical lasers such as ytterbium doped fiber laser, which is highly useful for high power and parametric applications.

This device can be also used in cascade configuration for obtaining a higher ratio of combining. For example, we used two units of the 4 × 1 power combiner and connected them to a 2 × 1 combiner, in order to obtain an 8 × 1 power combiner. Figure 7a shows the transmission of eight Gaussian sources at a 1 μm wavelength at $z = 0$, with a normalized power value of 0.125. Figure 7b shows the combining of eight sources to two at $z = 2.6$ mm. Figure 7c shows that eight Gaussian sources are combined to one single mode at $z = 5.8$ mm.

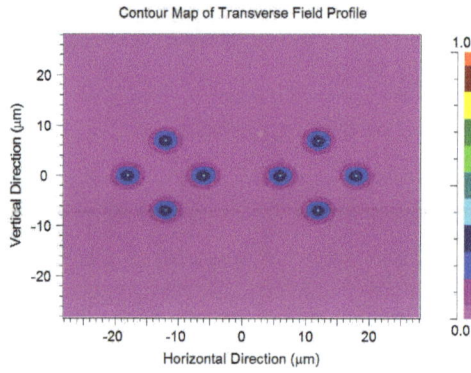

(a)

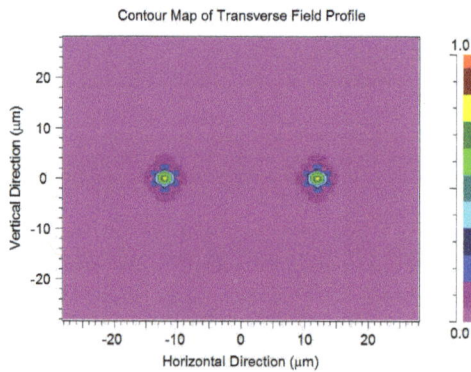

(b)

**Figure 7.** *Cont.*

Contour Map of Transverse Field Profile

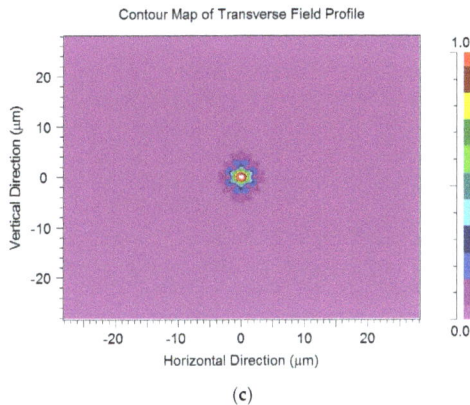

(c)

**Figure 7.** The 8 × 1 power combiner of the optical signals (λ = 1 μm) of three *xy* cross sections: (**a**) z = 0 mm; (**b**) z = 2.6 mm; (**c**) z = 5.8 mm.

## 5. Conclusions

In this article, we have shown a new approach of designing specific air holes in a multicore PCF structure that are utilized to obtain an optical 4 × 1 power combiner.

Through BPM simulation results, we showed that the energies of four optical signals at 1 μm wavelength can be combined into one source output in a PCF structure with dimensions of 38 μm × 38 μm × 2.6 mm. The amount of total power at the output is about 99.6% of the input power, the coupling losses between core 1–core 2/3/4/5 is about 0.01% of the input power, and the optical bandwidth value is 220 nm for the output port. Thus, this combiner device can be used in high power and parametric applications.

This combiner can be fabricated by stacking and drawing techniques which are simpler than other known fabrication techniques [20]. It is important to emphasize that all of these combiners were demonstrated using simulation.

In addition, a higher level 8 × 1 combiner was demonstrated using two units of 4 × 1 connected to a 2 × 1 combiner in a cascade configuration.

This research could be expanded to optical combiners with larger numbers of input fiber laser sources, such as 16 × 1 and 32 × 1.

**Author Contributions:** Dror Malka envisioned the project. Zeev Zalevsky provided guidance and funding. Dror Malka designed the device and performed simulations with support of Eyal Cohen, Dror Malka, Eyal Cohen, and Zeev Zalevsky wrote the manuscript text. Dror Malka made the figures and all authors reviewed the manuscripts.

**Conflicts of Interest:** The authors declare no conflict of interest.

## References

1. Snitzer, E. Optical maser action of $Nd^{+3}$ in a barium crown glass. *Phys. Rev. Lett.* **1961**, 7, 444–446. [CrossRef]
2. Snitzer, E.; Hoffman, F.; Crevier, R. Neodymium-glass-fiber laser. *J. Opt. Soc. Am.* **1963**, 53, 515–517.
3. Je, G.; Malka, D.; Kim, H.; Hong, S.; Shin, B. A study on micro hydroforming using shock wave of 355 nm UV-pulsed laser. *Appl. Surf. Sci.* **2017**, 417, 244–249. [CrossRef]
4. Ushenko, Y.A.; Arkhelyuk, A.D.; Sidor, M.I.; Bachynskyi, V.T.; Wanchuliak, O.Y. Laser polarization autofluorescence of endogenous porphyrins of optically anisotropic biological tissues and fluids in diagnostics of necrotic and pathological changes of human organs. *Appl. Opt.* **2014**, 53, B181–B191. [CrossRef] [PubMed]

5.  Ding, D.; Lv, X.; Chen, X.; Wang, F.; Zhang, J.; Che, K. Tunable high-power blue external cavity semiconductor laser. *Opt. Laser Technol.* **2017**, *94*, 1–5. [CrossRef]
6.  Jones, T.B.; Otterstrom, N.; Jackson, J.; Archibald, J.; Durfee, D.S. Laser wavelength metrology with color sensor chips. *Opt. Express* **2015**, *23*, 32471–32480. [CrossRef] [PubMed]
7.  Wang, Y.; Wang, Y.; Le, H.Q. Multi-spectral mid-infrared laser stand-off imaging. *Opt. Express* **2005**, *13*, 6572–6586. [CrossRef] [PubMed]
8.  Yahel, E.; Hardy, A. Modeling High-Power $Er^{3+}$-$Yb^{3+}$ Codoped Fiber Lasers. *J. Lightwave Technol.* **2003**, *21*, 2044. [CrossRef]
9.  Vienne, G.G.; Caplen, J.E.; Dong, L.; Minelly, J.D.; Nilsson, J.; Payne, D.N. Fabrication and Characterization of $Yb^{3+}$: $Er^{3+}$ Phosphosilicate Fibers for Lasers. *J. Lightwave Technol.* **1998**, *16*, 1990. [CrossRef]
10. Stiles, E. New developments in IPG fiber laser technology. In Proceedings of the 5th International Workshop on Fiber Lasers, Dresden, Germany, 30 September–1 October 2009.
11. Dawson, J.W.; Messerly, J.M.; Beach, R.J.; Shverdin, M.Y.; Stappaerts, E.A.; Sridharan, A.K.; Pax, P.H.; Heebner, J.E.; Siders, C.W.; Barty, C.P.J. Analysis of the scalability of diffraction-limited fiber lasers and amplifiers. *Opt. Express* **2008**, *16*, 13240–13266. [CrossRef]
12. Liu, Z.; Zhou, P.; Xu, X.; Wang, X.; Ma, Y. Coherent beam combining of high power fiber lasers: Progress and prospect. *Sci. China Technol. Sci.* **2013**, *56*. [CrossRef]
13. Drachenberg, D.; Divliansky, I.; Smirnov, V.; Venus, G.; Glebov, L. High Power Spectral Beam Combining of Fiber Lasers with Ultra High Spectral Density by Thermal Tuning of Volume Bragg Gratings. In *SPIE: Fiber Lasers VIII: Technology, Systems, and Applications, San Francisco, CA, USA, 22 January 2011*; SPIE: Bellingham, WA, USA, 2011; Volume 7914, p. 79141F.
14. Shamir, Y.; Zuitlin, R.; Sintov, Y.; Shtaif, M. High brightness efficient beam combining of 3 kW 1.07 µm fiber lasers with very low thermal dissipation. In Proceedings of the OASIS, Meeting on Optical Engineering and Science, Tel Aviv, Israel, 19–20 February 2013.
15. Russell, P.S.J. Photonic-crystal fibers. *J. Lightwave Technol.* **2006**, *24*, 4729–4749. [CrossRef]
16. Broeng, J.; Mogilevstev, D.; Barkou, S.E.; Bjarklev, A. Photonic crystal fibers: A new class of optical waveguides. *Opt. Fiber Technol.* **1999**, *5*, 305–330. [CrossRef]
17. Elbaz, D.; Malka, D.; Zalevsky, Z. Photonic crystal fiber based $1 \times N$ intensity and wavelength splitters/couplers. *Electromagnetics* **2013**, *32*, 209–220. [CrossRef]
18. Malka, D.; Zalevsky, Z. Multicore Photonic Crystal Fiber Based $1 \times 8$ Two-Dimensional Intensity Splitters/Couplers. *Electromagnetics* **2013**, *33*, 413–420. [CrossRef]
19. Malka, D.; Peled, A. Power Splitting of $1 \times 16$ in Multicore Photonic Crystal Fibers. *Appl. Surf. Sci.* **2017**, *417*, 34–39. [CrossRef]
20. Malka, D.; Sintov, Y.; Zalevsky, Z. Fiber-laser monolithic coherent beam combiner based on multicore photonic crystal fiber. *Opt. Eng.* **2014**, *54*, 011007. [CrossRef]
21. Mortimore, D.B. Theory and fabrication of $4 \times 4$ single-mode fused optical fiber couplers. *Appl. Opt.* **1990**, *29*, 371–374. [CrossRef] [PubMed]
22. Kumar, A.; Varshney, R.K.; Sinha, R.K. Scalar modes and coupling characteristics of eight-port waveguide couplers. *J. Lightwave Technol.* **1989**, *7*, 293–296. [CrossRef]
23. Mortimore, D.B.; Arkwright, J.W. Monolithic wavelength-falt-tened $1 \times 7$ single-mode fused fiber couplers: Theory, fabrication and analysis. *Appl. Opt.* **1991**, *30*, 650–659. [CrossRef] [PubMed]
24. Haus, H.A.; Huang, W. Coupled-Mode Theory. *Proc. IEEE* **1991**, *79*, 1505–1517. [CrossRef]
25. Katz, O.; Malka, D. Design of novel SOI $1 \times 4$ optical power splitter using seven horizontally slotted waveguides. *Photonics Nanostruct. Fundam. Appl.* **2017**, *25*, 9–13. [CrossRef]

*applied*
*sciences*

MDPI

*Article*

# Fast Frequency Acquisition and Phase Locking of Nonplanar Ring Oscillators

Yunxiang Wang *, Chen Wang, Yangping Tao, Yang Liu, Qiang Zhou , Jun Su, Zhiyong Wang, Shuangjin Shi and Qi Qiu

School of Optoelectronic Information, University of Electronic Science and Technology of China, Chengdu 610054, China; wsc868686@sina.com (C.W.); tyangping@163.com (Y.T.); 201621050224@std.uestc.edu.cn (Y.L.); betterchou@gmail.com (Q.Z.); stevensu27@126.com (J.S.); zywang@uestc.edu.cn (Z.W.); sjshi@uestc.edu.cn (S.S.); qqiu@uestc.edu.cn (Q.Q.)
* Correspondence: wangyunxiang@uestc.edu.cn; Tel.: +86-028-8320-1111

Academic Editor: Federico Pirzio
Received: 26 August 2017; Accepted: 4 October 2017; Published: 9 October 2017

**Abstract:** Optical phase locking is a critical technique in space coherent optical communication and active coherent laser beam combining. In a typical optical phase locking loop based on nonplanar ring oscillators, the pull-in range is normally less than 1 MHz, limited by loop delay and frequency tuning bandwidth of the laser source. Phase locking cannot be achieved at large initial frequency differences. In this work, a fast laser frequency acquisition method is demonstrated. The frequency difference between the signal and local lasers was measured via frequency dividing and period counting, and the frequency control signal was generated by a frequency discrimination and control module, to reduce the frequency difference to the pull-in range of the loop. Under the coordinating function of the loop filter and the frequency discrimination and control module, phase locking under a large initial frequency difference was achieved. The frequency acquisition range reached 164 MHz, and the acquisition and locking time was measured to be 440 ms. Additionally, the acquisition time was shortened with the decrease in initial frequency difference.

**Keywords:** optical phase-locking loop; frequency acquisition; coherent optical communication; frequency discrimination

---

## 1. Introduction

An optical phase locking loop (OPLL) plays an important role in space homodyne coherent optical communication systems, which have a significant advantage in receiving sensitivity. Gbps-level LEO-LEO [1], LEO-GEO [2], LEO-GS [3], and GEO-GS [4] communication links were verified, where LEO, GEO and GS denoted low earth orbit satellites, geostationary earth orbit satellite, and ground station, respectively. Compared to electrical wave sources in an electrical PLL, laser sources have a wider line width and a faster frequency drift [5], which result in difficult frequency acquisition and poor tracking performance. Early OPLLs were based on gas laser sources. A. L. Acholtz et al. achieved phase locking of $CO_2$ lasers and 100 Mbps homodyne communication [6]. However, limited by certain disadvantages in lifespan, volume, and efficiency, it is hard for gas lasers to meet certain requirements in space applications. The development of optical communication has greatly promoted the progress of semiconductor laser technology. S. Norimatsu et al. achieved phase locking of an external cavity semiconductor laser with a residual phase error of 7.4° [7]. Although semiconductor lasers have a small size and low cost, the linewidth of these devices is relatively wide, or the phase noise is large, which leads to losing lock or large residual phase error. Recently, integrated OPLLs based on semiconductor lasers have been studied intensively. Such loops have potential applications in optical frequency synthesis [8] and microwave photonics [9]. Integration of OPLLs reduces the loop delay, broadens the loop bandwidth, and improves the loop stability [10]. Thanks to the intensive study of diode-pumped solid-state lasers (DPSSL) over the last three decades, their performance has greatly improved in coherence, service life, and reliability. Among DPSSLs, nonplanar ring oscillators (NPROs) have a narrow linewidth (<1 kHz), low intensity noise (relative intensity noise <−140 dB/Hz), and good frequency stability. Due to the merit of a narrow linewidth, in OPLL applications, small residual phase error can be obtained. In homodyne coherent optical communication systems, small residual phase error is conducive to an improvement in the receiving sensitivity, which is important for space communication systems. Until now, NPRO is the only laser source that has been verified in orbit for applications of homodyne coherent optical communication, to the best of our knowledge. We believe that NPRO is, up to now, the first-choice laser source in space homodyne coherent optical communication systems.

For OPLLs based on NPRO, F. Herzog presented the scheme and performance of a dither loop, which did not require residual carrier transmission [11]. T. Ando et al. studied phase locking stability of OPLLs based on NPRO [12]. The authors analyzed the effect of response bandwidth of frequency tuning on phase locking precision [13]. Frequency acquisition performance is also quite important for phase locking loops. Frequency acquisition schemes based on a bang-bang phase detector [14] and a Hogge phase detector [15] have been reported in electrical PLLs. The range of initial frequency difference (or offset) between signal laser and local laser is larger in OPLLs, compared with electrical PLLs, and this is mainly due to laser frequency drift. Even under the condition of precise temperature control, the initial frequency offset can still be around 100 MHz. In addition, for OPLLs based on NPROs, the response bandwidth of the frequency tuning of the laser source is narrower, and loop delay is much longer, compared to electrical PLLs. Therefore, the lock-in range and pull-in range of an OPLL are much narrower. The frequency acquisition range, for OPLLs based on NPROs, should start from tens of kHz to more than 100 MHz. Thus, a special acquisition scheme with a wide range, high speed, and good compatibility with loop filters should be proposed for OPLLs based on NPROs. The frequency sweeping technique based on laser crystal temperature turning was used, with a frequency acquisition time longer than 10 s [2]. For this technique, if the loop lost locking due to a small frequency step during normal operation, it might also need more than 10 s for relocking. As the link duration of inter-satellite communication is typically in the range of 100–200 s for LEO-to-LEO links [1], the acquisition time should be reduced significantly to less than a few seconds, less than 1 s ideally. A shorter acquisition time means a longer communication time and more collected data. In this work, a faster frequency acquisition was achieved via frequency discrimination and piezoelectric frequency control. In addition, smooth continuity between the frequency acquisition process and the phase locking process was achieved. The frequency acquisition range reached 164 MHz, and the acquisition and locking time is less than 0.5 s.

## 2. Design of the OPLL

A block diagram of the OPLL is shown in Figure 1. The signal and local lasers were both NPROs, with two signal input ports for temperature and piezoelectric-transducer (PZT) frequency tuning, respectively. The temperature of the laser crystals was precisely controlled via a proportional–integral–derivative (PID) controller to keep the laser frequency stable. The control precision was 0.01 K for 1 h and 0.03 K for 8 h. The thermal tuning coefficient of Nd:YAG NPRO was 3 GHz/K. The typical frequency drift was measured to be around 30 MHz for 1 h and 100 MHz for 8 h. A piezoelectric element was bonded to the upper nonoptical face of the Nd:YAG crystal. The applied voltage can slightly change the dimension and refraction index of the laser crystal. Thus, the laser frequency was tuned. The PZT tuning coefficient was 1.6 MHz/V, and the response bandwidth was about 100 kHz.

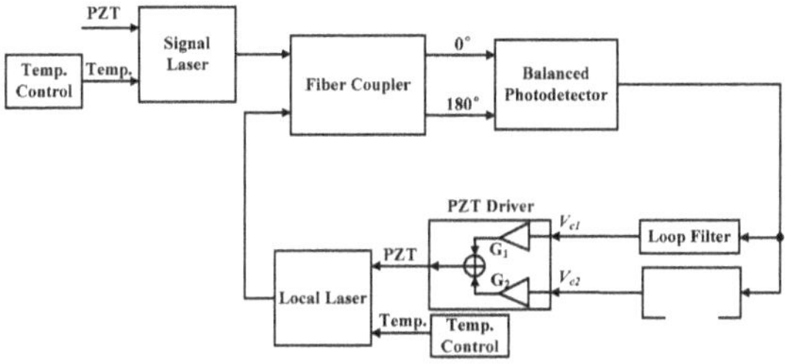

**Figure 1.** Block diagram of the optical frequency acquisition and the phase-locking loop.

The output lights of local and signal lasers were combined and mixed via a 50/50 fiber coupler. Two output beat signals with a 180° phase difference were sent to the balanced photodetector, which had a model number of PDB480C and was made by Thorlabs, Inc. (Newton, NJ, USA). The converted electrical beat signal, acting as a phase error signal, was sent to the loop filter and the frequency discrimination and control module (FDCM). $V_{c1}$ and $V_{c2}$ are the output signal amplitude of the loop filter and FDCM, respectively. The loop filter was an active proportional-plus-integral filter, where the first time constant $\tau_1 = 3$ ms and the second time constant $\tau_2 = 5$ μs. The main function of the loop filter was locking the phase, as the beat frequency falls into the pull-in range of the loop, which was around 700 kHz. The FDCM was used for the acquisition of signal frequency to keep the beat frequency below the pull-in range. $V_{c1}$ and $V_{c2}$ were amplified and added in the PZT driver, where the gain values $G_1$ and $G_2$ were 9 and 18, respectively. The output voltage amplitude of the PZT driver was ±190 V.

### 3. Principle of Frequency Acquisition

In the FDCM, the beat signal frequency is measured digitally, and a control signal is generated for frequency acquisition. A block diagram of the FDCM is shown in Figure 2. The frequency of the beat signal is firstly divided by 200 and converted into square pulses. The dividable frequency ranges from 10 kHz to 300 MHz. The pulse repetition frequency is measured via a microprogrammed control unit (MCU). The measuring period is set to be 5 ms. From the measured frequency in the current measuring period and the previous period, the frequency control direction and the frequency shift amount were obtained. An MCU generates corresponding binary data to the digital/analog (D/A) converter, in which the frequency control signal is generated. After passing through the amplifier, the control signal is sent to a low pass filter. It is then sent to the $V_{c2}$ port of the PZT driver. The low pass filter is essential for stable phase locking. In the locking status of the OPLL, the phase step and minor frequency step can cause a transient response in the phase error signal. Some frequency steps might be caused by the high frequency components of the D/A-converted signal. If the amplitude of the transient response signal reaches the threshold of the FDCM, it will function and generate a small frequency shifting signal. This is a false shifting signal and might result in losing lock. With a low pass filter of a 400 Hz bandwidth, the shifting signal is smoothed. The phase locking loop can precisely track the smoothed minor frequency shift, and keep a stable residual phase error. One frequency control cycle includes a frequency measuring period, 5 ms, and a frequency shifting period, 1 ms. The step response time of the low pass filter is close to 1 ms. Thus, the frequency shifting period is set to 1 ms.

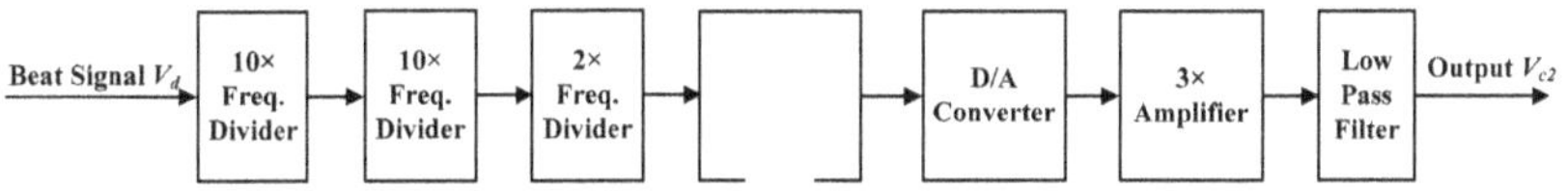

**Figure 2.** Block diagram of the frequency discrimination and control module (FDCM).

Figure 3 shows the flow chart of the frequency discrimination and control program. The program is mainly achieved via an MCU. After program initialization, pulses that are sent to the module during the measuring period are counted. The frequency difference $\Delta f_p$ in the current measuring period is computed from the counted pulse number. If $\Delta f_p$ is less than critical frequency difference $\Delta f_c$, then the frequency locking is achieved. Thus, the FDCM will maintain the frequency control voltage, and the loop will be locked under the function of the loop filter. The lock-in range and pull-in range of the loop are 15 kHz and 700 kHz, respectively. The critical frequency difference $\Delta f_c$ can be set to 40 kHz, with a locking time less than 1 ms under the function of the loop filter. If $\Delta f_p$ is larger than the frequency difference in the last measuring period, $\Delta f_b$, then the frequency control direction is incorrect. If the current frequency control cycle is the second cycle, the wrong control direction is caused by a random chosen direction in the first frequency control cycle. Otherwise, the wrong control direction is caused by fast laser frequency drift during the last frequency control cycle. The frequency control direction will thus be changed, and the frequency shift amount is determined by $\Delta f_p$ and the PZT tuning coefficient. If $\Delta f_p$ is less than $\Delta f_b$, then the frequency control direction is correct. The local laser frequency will be controlled in the correct direction until $\Delta f_p$ is less than $\Delta f_c$. Phase locking will thus be achieved via the loop filter.

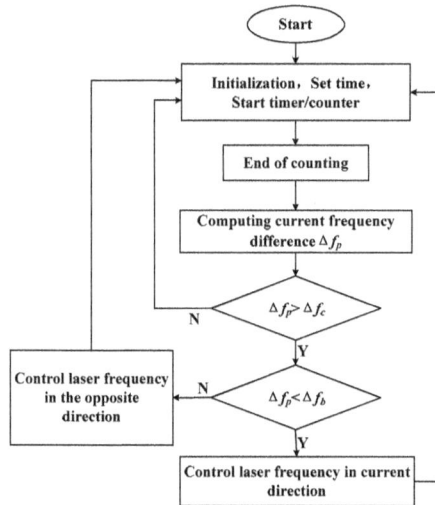

**Figure 3.** Flow chart of frequency discrimination and control program.

The choice of division ratio is a tradeoff between the maximum measurable frequency (MMF) and measurement accuracy (MA). The crystal oscillating frequency of the MCU is 24 MHz, corresponding to a machine cycle of 0.5 µs. For a division ratio of 200, MMF is 200 MHz, and MA is 40 kHz. For a smaller division ratio, to get the same MA, a shorter frequency measuring period can be adopted, and a faster acquisition can be achieved. However, with the MCU used in this work, the MA is increased. At the initial frequency difference that equals MA, the phase locking time is longer under the function of the loop filter. If the phase locking time is longer than one frequency control cycle, the OPLL will become unstable. For example, if the division ratio is set at 100, a frequency measuring period can be set at 2.5 ms, and one frequency control cycle is 3.5 ms, with an MA of 80 kHz. At an initial frequency difference of 80 kHz, the phase locking time is close to 3.5 ms, and the loop operates at the critical region with poor stability. Moreover, for a higher division ratio, the MMF is decreased, which is also undesired. Thus, the optimum division ratio in this work is 200. If we could decrease the phase locking time, under the condition of a shorter loop delay and a wider loop bandwidth, a smaller division ratio is allowed. The output voltage range of the D/A converter is ±5 V. The converted signal has a total amplification of 54× before sending to the PZT port of the local laser.

## 4. Experimental Results

The phase locking loop was constructed according to Figure 1. First, the FDCM was turned off, and the initial frequency difference of the two lasers was controlled by adjusting the crystal temperature of the signal laser. At a set frequency difference, as the FDCM was turned on, frequency acquisition and phase locking were tested. In the experiments, the output signal of the loop filter, the PZT driving signal, and the laser beat signal were monitored.

The frequency acquisition and phase locking process at an 8 MHz initial frequency difference is shown in Figure 4 with a time scale of 200 ms/div. At 630 ms, the FDCM was turned on, and the feedback control was started. After 80 ms, phase locking was achieved, as indicated by the beat signal (yellow). According to the frequency discrimination and control program, under the control of the FDCM, the local laser generated an initial frequency shift. As the change in frequency difference is measured, the direction of the frequency control and the frequency shift amount were determined. The main function of the initial frequency shift determined the right frequency control direction. After that, the frequency difference decreased to less than $\Delta f_c$, 40 kHz, as indicated by the PZT driving signal (green). For such a small frequency difference, the phase locking time based on the loop filter was quite short, shorter than 1 ms, which is negligible compared to the frequency acquisition time.

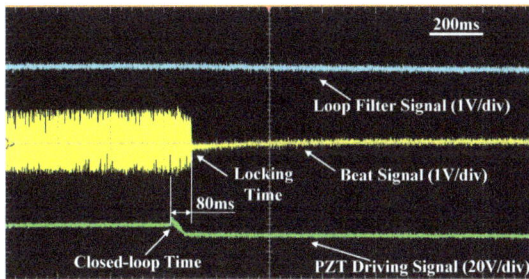

**Figure 4.** Frequency acquisition and phase locking process at an 8 MHz initial frequency difference.

As the initial frequency difference increased to 164 MHz, the acquisition time was around 440 ms, as shown in Figure 5. As the beat signal was detected via the probe of the oscilloscope, the amplitude of the detected signal varied with the signal frequency in the acquisition process because impedance matching varied at different frequencies. For a higher initial frequency difference, the frequency discrimination and control would go wrong sometimes. The main reason for this is that the machine cycle of the MCU, compared to the divided beat signal period, was so long that the frequency could not be correctly measured. It could be improved by increasing the crystal frequency and the processing speed of the MCU. In order to guarantee the stability of the frequency control process, the local laser frequency shift was set to less than 3 MHz at every control step. For an even higher frequency shift step, the laser frequency would not reach a steady state at the end of the control period, causing inaccurate frequency measuring in the following measuring period, even leading to false control. As the initial frequency difference increased, more control periods were needed, and acquisition time was normally extended. In the frequency acquisition process, when the beat signal frequency was out of the pull-in range of the loop, the loop filter output a constant DC signal. Only as the beat frequency fell into the pull-in range of the loop did the loop filter signal have an obvious response, as shown in the figure by the blue line. According to the measurement of the laser frequency stability, the frequency drift of the beat signal of the two lasers was around 100 MHz in 8 h, under precise temperature control.

The frequency acquisition range, 164 MHz, was significantly larger than the frequency drift range, so a stable loop operation was achievable.

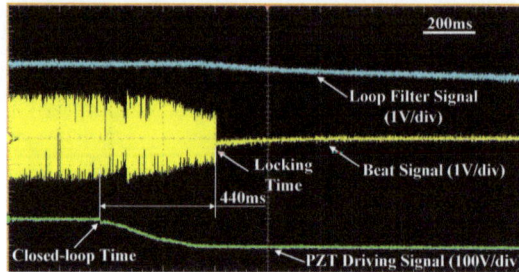

**Figure 5.** Frequency acquisition and phase locking process at a 164 MHz initial frequency difference.

The frequency acquisition and phase locking time at different initial beat frequencies is shown in Figure 6. According to the frequency discrimination and control program, the theoretical minimum number of control cycles was equal to the initial frequency difference divided by the maximum frequency shift step of a given period, which was set to 3 MHz. As the time of one control cycle was set to 6 ms, minimum acquisition time could be computed. The actual acquisition time was usually longer, and had a certain degree of uncertainty. Main causes include uncertainty of the initial frequency shift direction, laser frequency drift during the acquisition process, and a shortened frequency shift step in the last few control cycles. The dashed line in Figure 6 is the linear fitting line with a slope of 2.03 ms/MHz. The slope corresponds to the ratio of time of one frequency control cycle, 6 ms, to the maximum frequency shift step in one cycle, 3 MHz. Compared to the frequency acquisition time achieved via the frequency sweeping technique based on laser crystal temperature turning, which was longer than 10 s [2], the acquisition time of this work was 1–2 orders of magnitude shorter.

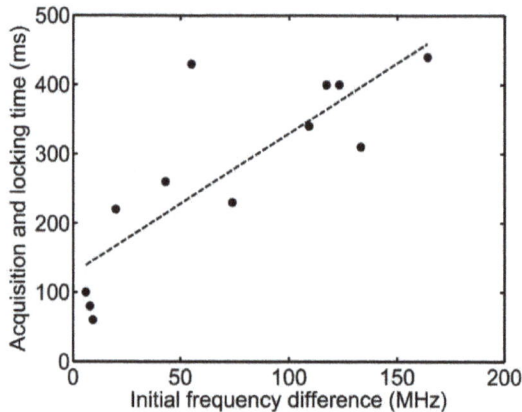

**Figure 6.** Measured frequency acquisition and phase locking time vs. the initial frequency difference. The dashed line is the linear fitting line.

## 5. Conclusions

Nonplanar ring oscillators have a high frequency stability, a narrow linewidth, and low intensity noise. These devices are becoming the preferred laser sources in coherent space optical communications. They can also be used in coherent laser beam combinations. In these applications, phase locking is the key technique. Traditionally, laser frequency acquisition based on temperature control consumes quite a long time. In this work, we designed and achieved a frequency acquisition sub-loop based on PZT control. The acquisition sub-loop operated coordinately with the phase locking loop based on the loop filter. The frequency acquisition range reached 164 MHz, with an acquisition time of less than 500 ms. For a smaller initial frequency difference, acquisition times decreased. This technique significantly shortens the frequency acquisition time, which makes the system reach work status much more quickly. In addition, by improving the processing speed of the control circuit, acquisition time can be further shortened.

**Acknowledgments:** This research is supported by the National Natural Science Foundation of China (Grant No. 61308041 and 61405030) and the PetroChina Innovation Foundation (Grant No. 2017D-5007-0603).

**Author Contributions:** Yunxiang Wang and Qi Qiu conceived and designed the experiments, Yangping Tao and Yang Liu designed the feedback circuits; Qiang Zhou and Zhiyong Wang performed the experiments; Jun Su and Shuangjin Shi analyzed the data; Chen Wang designed the nonplanar ring oscillators.

**Conflicts of Interest:** The authors declare no conflict of interest.

## References

1. Smutny, B.; Kaempfner, H.; Muehlnikel, G.; Sterr, U.; Wandernoth, B.; Heine, F.; Hildebrand, U.; Dallmann, D.; Reinhardt, M.; Freier, A.; et al. 5.6 Gbps Optical Intersatellite Communication Link. In Proceedings of the SPIE—The International Society for Optical Engineering, San Jose, CA, USA, 24 February 2009; No. 7199. pp. 601–608.
2. Tröndle, D.; Pimentel, P.M.; Rochow, C.; Zech, H.; Muehlnikel, G.; Heine, F.; Meyer, R.; Philipp-May, S.; Lutzer, M.; Benzi, E.; et al. Alphasat-Sentinel-1A Optical Inter-Satellite Links: Run-Up For The European Data Relay Satellite System. In Proceedings of the SPIE—The International Society for Optical Engineering, San Francisco, CA, USA, 15 March 2016; No. 9739. pp. 201–206.
3. Gregory, M.; Troendle, D.; Muehlnikel, G.; Heine, F.; Meyer, R.; Lutzer, M.; Czichy, R. 3 Years Coherent Space to Ground Links: Performance Results And Outlook For The Optical Ground Station Equipped with Adaptive Optics. In Proceedings of the SPIE—The International Society for Optical Engineering, San Francisco, CA, USA, 19 March 2013; No. 8610. pp. 401–413.
4. Saucke, K.; Seiter, C.; Heine, F.; Gregory, M.; Tröndle, D.; Fischer, E.; Berkefeld, T.; Feriencik, M.; Feriencik, M.; Richter, I.; et al. The Tesat Transportable Adaptive Optical Ground Station. In Proceedings of the SPIE—The International Society for Optical Engineering, San Francisco, CA, USA, 15 March 2016; No. 9739. pp. 601–611.
5. Kang, H.; Zhang, H.; Wang, D. Thermal-induced refractive-index planar waveguide laser. *Appl. Phys. Lett.* **2009**, *95*, 181102.
6. Scholtz, A.L.; Leeb, W.R.; Philipp, H.K.; Bonek, E. Infra-Red Homodyne Receiver with Acousto-Optically Controlled Local Oscillator. *Electron. Lett.* **1983**, *19*, 234–235. [CrossRef]
7. Norimatsu, S.; Iwashita, K.; Noguchi, K. 10 Gbit/s Optical PSK Homodyne Transmission Experiments Using External Cavity DFB LDs. *Electron. Lett.* **1990**, *26*, 648–649. [CrossRef]
8. Arafin, S.; Simsek, A.; Kim, S.; Dwivedi, S.; Liang, W.; Eliyahu, D.; Klamkin, J.; Matsko, A.; Johansson, L.; Maleki, L.; et al. Towards Chip-Scale Optical Frequency Synthesis Based on Optical Heterodyne Phase-Locked Loop. *Opt. Express* **2017**, *25*, 681–695. [CrossRef] [PubMed]
9. Balakier, K.; Fice, M.J.; Ponnampalam, L.; Seeds, A.J.; Renaud, C.C. Monolithically Integrated Optical Phase Lock Loop for Microwave Photonics. *J. Lightwave Technol.* **2014**, *32*, 3893–3900. [CrossRef]
10. Lu, M.; Park, H.; Bloch, E.; Johansson, L.A.; Rodwell, M.J.; Coldren, L.A. An Integrated Heterodyne Optical Phase-Locked Loop with Record Offset Locking Frequency. In Proceedings of the 2014 Optical Fiber Communication Conference and Exhibition (OFC), San Francisco, CA, USA, 9–13 March 2014.
11. Herzog, F. An Optical Phase Locked Loop for Coherent Space Communications. Ph.D. Thesis, Swiss Federal Institute of Technology, Zurich, Switzerland, 2006.
12. Ando, T.; Haraguchi, E.; Tajima, K.; Hirano, Y.; Hanada, T.; Yamakawa, S. Optical Homodyne BPSK Receiver with Doppler Shift Compensation for LEO-GEO Optical Communication. In Proceedings of the 2013 Conference on Lasers and Electro-Optics Pacifc Rim, Kyoto, Japan, 30 June–4 July 2013.

13. Wang, Y.; Qiu, Q.; Shi, S.; Su, J.; Liao, Y.; Xiong, C. High-precision optical phase locking based on wideband acousto-optical frequency shifting. *Chin. Opt. Lett.* **2014**, *12*, 021402. [CrossRef]

14. Shu, G.; Choi, W.S.; Saxena, S.; Anand, T.; Elshazly, A.; Hanumolu, P.K. 8.7 A 4-to-10.5 Gb/s 2.2 mW/Gb/s continuous-rate digital CDR with automatic frequency acquisition in 65 nm CMOS. In Proceedings of the 2014 IEEE Interational Solid-State Circuits Conference Digest of Technical Papers (ISSCC), San Francisco, CA, USA, 9–13 February 2014; pp. 150–151.

15. Huang, S.; Cao, J.; Green, M.M. An 8.2 Gb/s-to-10.3 Gb/s Full-Rate Linear Referenceless CDR Without Frequency Detector in 0.18 μm CMOS. *IEEE J. Solid State Circuits* **2015**, *50*, 2048–2060. [CrossRef]

*applied
sciences*

MDPI

*Article*

# Effect of Polishing-Induced Subsurface Impurity Defects on Laser Damage Resistance of Fused Silica Optics and Their Removal with HF Acid Etching

**Jian Cheng [1,\*], Jinghe Wang [1], Jing Hou [2], Hongxiang Wang [1] and Lei Zhang [1]**

[1]  School of Mechatronics Engineering, Harbin Institute of Technology, Harbin 150001, China;
    hitwjh@163.com (J.W.); whx@hit.edu.cn (H.W.); emailzhanglei@126.com (L.Z.)
[2]  Research Center of Laser Fusion, China Academy of Engineering Physics, Mianyang 621900, China;
    houjing1997@163.com
\*  Correspondence: cheng.826@hit.edu.cn; Tel.: +86-451-8640-3252

Academic Editor: Federico Pirzio
Received: 11 July 2017; Accepted: 11 August 2017; Published: 15 August 2017

**Abstract:** Laser-induced damage on fused silica optics remains a major issue that limits the promotion of energy output of large laser systems. Subsurface impurity defects inevitably introduced in the practical polishing process incur strong thermal absorption for incident lasers, seriously lowering the laser-induced damage threshold (LIDT). Here, we simulate the temperature and thermal stress distributions involved in the laser irradiation process to investigate the effect of impurity defects on laser damage resistance. Then, HF-based etchants (HF:NH$_4$F) are applied to remove the subsurface impurity defects and the surface quality, impurity contents and laser damage resistance of etched silica surfaces are tested. The results indicate that the presence of impurity defects could induce a dramatic rise of local temperature and thermal stress. The maximum temperature and stress can reach up to 7073 K and 8739 MPa, respectively, far higher than the melting point and compressive strength of fused silica, resulting in serious laser damage. The effect of impurity defects on laser damage resistance is dependent on the species, size and spatial location of the defects, and CeO$_2$ defects play a dominant role in lowering the LIDT, followed by Fe and Al defects. CeO$_2$ defects with radius of 0.3 μm, which reside 0.15 μm beneath the surface, are the most dangerous defects for incurring laser damage. By HF acid etching, the negative effect of impurity defects on laser damage resistance could be effectively mitigated. It is validated that with HF acid etching, the number of dangerous CeO$_2$ defects is decreased by more than half, and the LIDT could be improved to 27.1 J/cm$^2$.

**Keywords:** fused silica; ultra-precision polishing; subsurface damage; laser damage resistance; absorbing impurity defects; HF acid etching

## 1. Introduction

In order to achieve clean and sustainable energy resources, high-power laser systems have been developed worldwide for pursuing inertial confinement fusion (ICF), such as the National Ignition Facility (NIF) in the United States [1,2], the Laser MegaJoule (LMJ) in France [3], the High Power laser Energy Research facility (HiPER) in Europe [4] and the ShenGuang (SG)-III laser facility in China [5]. To obtain the extremely high pressure and temperature required for ICF ignition, a huge amount of large-aperture optics with high precision surfaces are required to temporally, spatially and spectrally control the laser beams. The laser beams are finally coupled together, focusing simultaneously on a very tiny target filled with fusion fuels. Among these optics, fused silica is an amorphous state of silicon dioxide, which has extremely low thermal conductivity, super strong thermal-shock resistance, low dielectric loss, and high deformation (1370 K) and softening (2000 K) temperatures. Besides, the fused silica optics possess such broad optical transmission spectra that over 1000 pieces of silica optics have been widely used in the sub-laser systems with fundamental ($1\omega$) and triple ($3\omega$) frequencies of the ICF facilities. The fused silica optics serve as switch and vacuum windows, wedged focus lens, diffraction grating, debris shields and so on [6,7]. Since the fused silica optics are exposed to high-power lasers in their actual applications, the well prepared engineering silica components should have very high laser damage resistance. However, due to the weak mechanical properties of fused silica, the introduction of undesirable by-products (viewed as defects) on the surface or subsurface of finished brittle silica parts in the actual cutting, grinding, polishing, coating and cleaning processes is inevitable. The surface and subsurface defects can extend downward several to tens of microns beneath the finished surface, which would greatly lower the energy output capacity of the ICF laser facilities. Currently, thanks to the development of various new advanced processing techniques (e.g., laser conditioning [8], thermal annealing [9,10]), the bulk laser-induced damage threshold (LIDT) of fused silica can reach up to $475 \pm 25\,\mathrm{GW/cm^2}$ (1064 nm, 8 ns), which is very close to the theoretically intrinsic LIDT and almost one order of magnitude higher than the surface LIDT [11]. This means that the laser-induced damage on surface and subsurface of fused silica plays a dominant role in restricting the promotion of energy output capacity of ICF facilities. Similar to the fused silica optics applied in high power laser systems, many other optical materials (e.g., potassium dihydrogen phosphate KDP as well as its deuterated analog DKDP, fluoride and selenide crystals) are also expected to possess high optical qualities, like optical transparency and laser damage threshold [12–15]. For these crystal materials, rapid growth of large-size boules with desired optical properties has been being a great challenge, and the growth conditions are susceptible to affect the crystal structures, which is closely associated with the bulk laser damage threshold. As a result, the efforts to promote the optical properties of these optical materials are mainly focused on exploring the mechanisms of bulk laser-induced damage and optimizing the growth parameters to improve the crystal structures for increasing the bulk LIDT [14–16]. For fused silica optics, though the internal structure of silica is stable and robust, its hard and brittle characteristics would make the ultra-precision manufacturing of defect-free silica surfaces a great challenge. Therefore, the study on laser damage of fused silica optics should primarily aim at the defects on a finished surface or subsurface. It is of great theoretical and practical significance to explore the underlying mechanisms involved in the laser-induced surface damage and develop new engineering techniques to remove the surface and subsurface defects for improving the laser damage resistance of fused silica.

The primary source leading to the low surface LIDT of fused optics is the subsurface damage (SSD) caused in the chemical-mechanical polishing process. The SSD layer located beneath the polished surface generally consists of re-deposition layer (also named the Beilby layer), a defect layer with crack and scratch defects included, and a deformed layer [17–19]. During the polishing process, the highly absorptive impurities (e.g., Ce, Fe, etc.) coming from the polishing slurries are randomly distributed among the re-deposition layer. Some of the impurities are even embedded and hidden deeply inside the defect layer via entering the open subsurface cracks and scratches. Under the irradiation of intense laser, the impurity defects would make the natively transparent fused silica optics highly absorptive to incident laser, resulting in very high local temperature and stress, and eventually the breakdown of the optical parts [7,20,21]. Meanwhile, the absorptive impurity defects would change the initial band-gap structure of fused silica and also trigger new photon excitation under intense laser irradiation. These effects caused by the impurity defects would substantially affect the laser-induced nonlinear excitation of dielectric silica optics (e.g., multi-photon and avalanche ionizations), making the optical material more susceptible to laser damage [22,23]. The absorptive impurities mainly originate from three kinds of sources in the optical processing stage: oxide polishing slurries ($CeO_2$, $ZrO_2$, etc.), metal polishing tools (Fe, Cu, Cr, etc.) and Al ion from optical cleaning. Hu et al. [24] demonstrated that besides subsurface scratch and dig defects, ~35% of the laser damage sites were initiated at invisible defects, probably being absorptive impurities with submicrometer size. It was also experimentally proved by Neauport et al. [17,25] that the laser-induced damage on fused silica was closely associated with the subsurface absorptive impurities and different types of impurity defects would incur different levels of optical damage. Hence, in this work, the heat conduction and thermo-elastic equations involved in the absorbing process of impurity defects under intense laser irradiation are firstly numerically resolved using the finite element method (FEM). Then, the temperature and thermal stress distributions inside fused silica caused by impurity defects are investigated to figure out the most dangerous species and size range of impurity defects in decreasing the laser damage resistance of silica optics. This part of work could provide further understanding of the laser-induced damage mechanisms on optical components, which are beneficial to the surface/subsurface quality evaluation and SSD removal of ultra-precision fabricated fused silica optics.

To alleviate the effect of SSD defects on the laser damage resistance of fused silica optics, many engineering techniques (e.g., hydrofluoric HF acid etching [26–28], ultraviolet or $CO_2$ laser preprocessing [8,29], plasma etching [30], and magneto-rheological finishing (MRF) [31], etc.) have been developed and applied in the actual preparation processes of high-quality silica surfaces. In the laser preprocessing process, lasers with energy lower than the LIDT are used to irradiate the fused silica surfaces. Due to the strong absorption of silica material to $CO_2$ lasers, the temperature on the optical surface would moderately rise, resulting in the recombination of surface micro-structure and the healing of subsurface cracks. At the same time, the lasers can bring about the electronic excitation of the impurity defects, which would finally lower the defect energy band to a steady state. Hence, this processing technique is able to mitigate the negative effect of impurity defects and subsurface cracks on the laser damage resistance of fused silica. However, there are some shortcomings that need to be urgently solved as well. For example, the temperature gradient in laser heated zone can produce high residual thermal stress and the surface figure error caused by laser preprocessing may induce wavefront distortion, which greatly limits the capacity and advancement of this technique in improving the laser damage resistance [32]. For the plasma etching process, carbon tetrafluoride ($CF_4$) gas is necessary for effectively etching the silica materials. However, the $CF_4$ gas and its residual etching reaction product may become new sources of impurity defects, polluting the processed surface and consequently influencing the effect of improving the laser damage resistance [30]. However, high-quality optical surfaces with low roughness and few SSD defects can be achieved with the MRF method, and the magnetic Fe-ion defects from the magnetorheological fluid may be deposited inside the finished optical surfaces, resulting in low LIDT. However, for the HF acid etching technique, polished fused silica optics are immerged in various HF-based etchants (HF or $NH_4F$:HF mixed solutions with various ratios) to remove the re-deposition layer and partial SSD defects. With the HF acid etching, the crack and scratch defects hidden in deep subsurface of fused silica can be blunted and no new impurity defects would be introduced as well [26]. Besides, under the assistance of megahertz-frequency agitation, the scouring effect of micro jet on the cleaned optical surfaces could be enhanced, causing the micro-particles and impurity defects adhering to the optical surface/subsurface to be further removed [28,33]. As a result, the laser damage resistance of finished fused silica would be greatly enhanced. In the latter part of this work, the HF acid etching technique is employed to remove the SSD defects on fused silica. Then, the etching rate, impurity contents, surface quality and laser damage resistance of etched silica samples are thoroughly investigated and analyzed to validate the role of the HF acid etching technique in removing the subsurface impurity defects and promoting the laser energy capacity of fused silica.

## 2. Theory and Methods

### 2.1. Modeling of Impurity-Induced Temperature and Thermal Distributions

The residual impurity defects are generally distributed inside the SSD layers of ground and polished fused silica optics. The laser energy absorbed by these impurity defects is far greater than that by intrinsic thermal absorption of the fused silica itself. On the one hand, the strong absorption of these impurity defects can locally heat the silica materials, inevitably resulting in unsteady and non-uniform temperature distribution. When the local temperature reaches a certain degree, the optical material may suffer from modification, softening, melting and even boiling, which would certainly incur laser-induced damage [7]. On the other hand, the great temperature gap inside silica materials caused by the absorption of impurity defects could give rise to internal thermal stress, and correspondingly leads to the negative effects of obvious wavefront distortion and weak mechanical strength. These negative effects would directly affect the laser energy capacity and service life of large-aperture fused silica optics [20,21]. In the present efforts, the heat transfer and thermo-elastic equations governing the laser energy absorbing process are numerically solved by adopting the FEM method to investigate the distributions of temperature and thermal stress induced by impurity defects. The effects of subsurface impurity defects with various species and dimensions on the laser damage resistance are then analyzed, based on which we finally clarify the most dangerous impurity defects in terms of lowering the LIDT of fused silica optics.

When the incident laser falls on the silica surface, most of the laser energy is reflected at the interface or transmitted through the bulk, leaving only a fraction of the laser energy absorbed by the materials [34]. As the absorbed energy penetrates, the temperature on fused silica materials increases. The spatially and temporally increasing temperature can be described by the heat transfer equation in the Cartesian coordinate system which is given below [35]:

$$\frac{\partial T}{\partial \tau} = \alpha \left( \frac{\partial^2 T}{\partial x^2} + \frac{\partial^2 T}{\partial y^2} + \frac{\partial^2 T}{\partial z^2} \right) + \frac{q_v}{\rho c} \tag{1}$$

where, $T$, $\alpha$, $c$ and $\rho$ represent the temperature, thermal diffusivity, specific heat capacity, and density, respectively. $q_v$ indicates the heat source determined by the incident laser, which can be expressed as [36]:

$$q_v = \alpha \frac{(1-R)P}{\pi a^2} \exp\left(-\frac{x^2+y^2}{a^2}\right) \exp(-\alpha z) \tag{2}$$

with $\alpha$ the absorption coefficient, $R$ the Fresnel reflection coefficient, $P$ the incident laser power, and $a$ the radius waist at $1/e$.

As the laser irradiates the optical surface, part of the heat energy is transported into the bulk, resulting in the rise of surface temperature. Thermal radiation would take place at the hot surface. Besides, the high temperature on the surface may cool down by exchanging heat energy with the ambient air in the form of thermal convection. Hence, the below boundary conditions of heat flux, natural convection cooling and surface-to-ambient radiation are all applied to fully describe the heat transfer process in fused silica under the intense laser irradiation [37]. Under the control of Equation (3), the energy balance would be actually realized by transferring the energy among the sample surface, sample bulk and the surrounding circumstance during the laser heating and natural cooling processes.

$$I\Delta\tau = -k\frac{\partial T}{\partial z} + h(T_s - T_0) + \sigma\varepsilon(T_s^4 - T_0^4) \tag{3}$$

where, $k$, $h$ and $\varepsilon$ are the heat conductivity, convection coefficient, and radiation coefficient, respectively. $T_s$ and $T_0$ denote the temperatures of fused silica surface and indoor air temperature. $I$ represents the incident laser intensity and $\Delta t$ indicates the laser action period. $\sigma$ is the Boltzmann constant.

The historical temperature evolution on fused silica optics can be obtained by numerically solving the heat transfer equation of Equation (1) with the consideration of boundary conditions of Equation (3). It is worth noting that in this work we only focus on the temperature evolution inside the laser spot for the reason that the amount of heat energy transferred outside the light spot is much smaller.

Under the irradiation of the Gaussian laser pulse, the thermal distortion caused by absorptive impurity defects could be resolved and analyzed on the basis of the temperature distribution obtained from Equation (1). Ignoring the volume and inertia forces inside fused silica, the thermal distortion arising from internal temperature gradient would be described as follows [38]:

$$\begin{cases} \nabla^2 u_r - \frac{u_r}{r^2} + \frac{1}{1-2v}\frac{\partial e}{\partial r} - \frac{2(1+v)}{1-2v}\alpha_t\frac{\partial T}{\partial r} = 0 \\ \nabla^2 u_z + \frac{1}{1-2v}\frac{\partial e}{\partial z} - \frac{2(1+v)}{1-2v}\alpha_t\frac{\partial T}{\partial z} = 0 \end{cases} \tag{4}$$

where $u_r$ and $u_z$ are the displacement components in $r$ and $z$ directions of cylindrical coordinate system. $e$ represents the volumetric strain of fused silica. $v$ and $\alpha_t$ are the Poisson's ratio and coefficient of thermal expansion, respectively. Using the relations between stress and strain, the expression of potential energy principle can be gained based on the principle of virtual work. Then, by dispersing the potential energy expression and solving the nodal displacement matrix with the FEM method, we would obtain the corresponding thermal stress distribution initiated by the absorptive impurity defects.

With the application of advanced testing techniques like synchrotron radiation X-ray fluorescence (SXRF) and secondary ion mass spectrometry (SIMS), it has been previously reported [17,18,20,22,25] that the precision grinding and polishing of optical materials can probably introduce impurity defects with sub-wavelength size on the processed surface or subsurface. The impurity defects mainly contain cerium oxide ($CeO_2$), zirconium dioxide ($ZrO_2$), iron (Fe), aluminum (Al), chromium (Cr) and so on [17,20,25]. In this work, we choose $CeO_2$, Fe and Al as the three representative defects coming from the oxide polishing slurries, metal polishing tools and cleaning solutions, respectively. The effects of defect species, size and spatial location are all investigated and compared to reveal the most dangerous impurity to laser damage resistance. The optical and thermodynamic parameters of fused silica and impurity defects are exhibited in Tables 1 and 2, respectively. It should be noted that the optical damage of fused silica materials initiated by impurity defects under intense laser irradiations is really a complex process, which involves in the material modification, softening, melting, boiling, material fracture and even ejection. All of these processes are closely associated with the rise of temperature and stress during the energy absorption of incident lasers. Therefore, in the present work, the material definition models for both fused silica and impurity defects are simplified with constant thermodynamic parameters to model the temperature and stress distributions caused by representative impurity defects for evaluating their effects on the laser damage resistance.

**Table 1.** Thermodynamic parameters of fused silica materials used in calculations [39,40].

| Property | Nomenclature | Value (Units) |
|---|---|---|
| Molar mass | $M$ | 60.06 (g/mol) |
| Crystal system | – | Amorphous |
| Density | $\rho$ | 2.21 (g/cm$^3$) |
| Melting point | $T_m$ | 1900 (°C) |
| Thermo-optical coefficient | $\varepsilon$ | $1 \times 10^{-5}$ |
| Coefficient of linear expansion | $\alpha_t$ | $5.5 \times 10^{-7}$ °C$^{-1}$ |
| Specific heat capacity | $c$ | 0.728 J/(g·°C) |
| Heat conductivity coefficient | $k$ | 1.35 W/(m·°C) |
| Relative dielectric constant | $\varepsilon_r$ | 2.25 |
| Young modulus | $E$ | $7.36 \times 10^{10}$ (Pa) |
| Shear modulus | $G$ | $3.14 \times 10^{10}$ (Pa) |
| Compressive strength | $P$ | 800~1000 (MPa) |
| Poisson's ratio | $v$ | 0.17 |

**Table 2.** Thermodynamic parameters of various impurity defects used in the calculation [40–42].

| Defect Species | Density (g/cm$^3$) | Specific Heat Capacity (J/(g·°C)) | Heat Conductivity (W/(cm·°C)) | Coefficient of Linear Expansion ($\times 10^{-6}$ K$^{-1}$) | Young Modulus (GPa) | Poisson's Ratio |
|---|---|---|---|---|---|---|
| Fe | 7.0 | 0.45 | 0.565 | 11.8 | 152.0 | 0.30 |
| Al | 2.7 | 0.88 | 2.38 | 23.0 | 70.0 | 0.33 |
| CeO$_2$ | 7.13 | 0.465 | 0.045 | 13.2 | 174 | 0.32 |

Figure 1 shows the FEM model for simulating the temperature and thermal stress distributions caused by subsurface impurity defect. To improve the computational efficiency, only the domain exposed within the laser spot is calculated and the diameter of fused silica is set to be 400 μm. The sample thickness is chosen as 10 μm due to the dramatically attenuating of laser energy in the vertical propagation direction. A two-dimensional axisymmetric finite element model is applied in this work for a cylindrical fused silica optics. The diabatic boundary conditions combined with the surface-to-ambient radiation boundary are applied at the borders of the simulation domain. Since the energy of the applied incident lasers is spatially distributed with a Gaussian profile, the temperature in the area beyond the laser spot would be much lower than those in the central area. Hence, the applied boundary conditions would not affect the simulation accuracy and more attention should be paid on the results of temperature and stress fields near the central area. The laser parameters chosen in the simulations are the same as those applied in the actual laser damage experiments to ensure accurate and reliable simulation results. It is worth noting that the actual laser energy is spatially distributed with an approximate Gaussian distribution. For simplicity, the spatial distribution of incident lasers with Gaussian shape (see Figure 1c) is applied in the simulations with a 390-μm beam diameter and a 10-ns pulse duration. The laser frequency is 1 Hz and after each laser pulse the fused silica sample cools down naturally. In this work, the evolutions of temperature and stress distributions during the laser heating and free cooling processes are both simulated and the total simulation running time is set to be much longer than the pulse duration to insure the stable temperature and stress fields to be reached finally. For simplicity, the impurity defect inside fused silica in Figure 1b is viewed as a small spherical particle positioned right beneath the laser spot center. The simulation domain is non-uniformly gridded in different regions: the defect region is gridded with triangular mesh, while the other region is gridded with quadrilateral mesh. Further, the refined gridding sizes are applied in the vicinity of fused silica-defect interface to guarantee the simulation accuracy. We have adjusted the maximum element mesh size to check the deviations of the simulation results. It has been found that the simulation results with the applied mesh sizes are maintained almost equal to those with more refined meshes, which indicates that the mesh generation is applicable to the simulations.

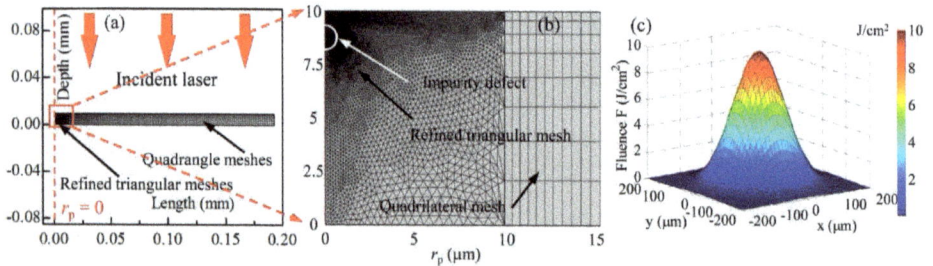

**Figure 1.** The finite element method (FEM) model for simulating the temperature and thermal stress distributions initiated by impurity defects under intense laser irradiation: (a) the FEM model; (b) the designed gridding; (c) the spatial energy distribution of applied Gaussian laser pulse.

### 2.2. Sample Preparation, SSD Removal, and Etching Process Characterization

In the practical chemical-mechanical polishing process of fused silica optics, mechanical and chemical actions are both involved in generating the polished surfaces by mutual contact, friction, squeezing and deformation between polishing agent and optical surface. Under the combined work of mechanical and chemical actions, the optical materials are removed, leaving the subsurface damage (SSD) distributed beneath the polished surface. The SSD distribution profile is presented in Figure 2, which includes re-deposition, defect and deformed layers [17,26,31,43]. The re-deposition layer consists of impurity defects and silica compounds, resulting from the hydrolysis reaction of fused silica.

The defect layer is mainly formed by cracks and scratches introduced in the initial rough grinding process. Besides, in the final polishing process, the impurity defects are easily imbedded and hidden inside the defect layer as well. The deformed layer is generated by the deformation of optical materials, which ascribes to the pressure stress of grinding wheel and buffing pad in the processing processes.

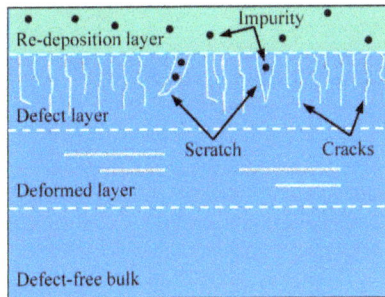

**Figure 2.** Schematic of the subsurface damage (SSD) defect distribution located beneath the ground and polished fused silica surface.

To promote the laser damage resistance of fused silica optics, the HF acid etching technique is employed in this work to remove the subsurface impurity defects. In the etching process, the topside re-deposition layer including most of the impurity defects is firstly dissolved by chemical reaction between the HF acid and hydrolysis products. Then, the crack and scratch defects are blunted and the embedded impurity can be also easily wiped off. In this way, the detrimental impact of these SSD defects on the laser damage resistance can be greatly mitigated for the silica optics. In the sample preparing process, W14-sized SiC abrasive particles are adopted to grind the ten pieces of initial fused silica samples (30-mm diameter and 5-mm thickness) for half an hour. The ground samples are then polished for 2 h with 1-$\mu$m-diameter $CeO_2$ polishing slurry to ensure that large brittle scratches and cracks are totally removed or buried under the re-deposition layer. The grinding and polishing spindle speeds are 50 r/min and 40 r/min. The grinding and polishing solution concentrations are 10 wt % and 8 wt %, respectively, with a 22-kPa load pressure. Figure 3 shows the HF acid etching schematic for removing the subsurface impurity defects. The HF-based etchant with 2% HF (mass fraction) and 5% $NH_4F$ (mass fraction) is applied with solvent of deionized water. The $NH_4F$ solvent, regarded as buffered oxide etch (BOE), is added in the etching process to promote the $F^-$ population and eventually guarantee the stable etching rate and low evaporation pressure. It should be noted that one half side of the topmost sample surface is set as reference with the paraffin applied over it to prevent the acid etching. Meanwhile, the other half side of the polished silica surface is etched when submerged into the HF-based etchant. After different periods of acid etching, the samples are cleaned with deionized water and the coated paraffin layers are removed with acetone reagent. The effectiveness of acid etching on removing the subsurface defects of polished silica optics is presented by the etching depths at various etching times, which can be directly calculated from a key parameter of etching rate. The etching rate in this work is measured with a stylus profilometer and the principle diagram of the measuring process is shown in Figure 3b. For silica samples etched for various periods, the profilometer stylus scans perpendicularly to the dividing line of the un-etched and etched surfaces to obtain the height profiles before and after HF acid etching. By comparing the height profiles of the two surface regions, the etching depths under various etching periods can be calculated, and the curve of etching rate is then accordingly achieved as well (see Figure 3b).

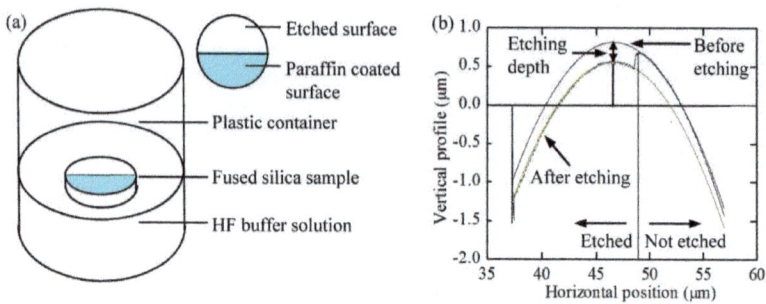

**Figure 3.** Sketch of the HF acid etching experiment (**a**) and the principle diagram; (**b**) for determining the etching depth using stylus profilometer in the etching process of fused silica optics.

Besides, in order to investigate the dynamic evolution of the surface quality involved in the etching process, the surface roughness and morphology are both tested at various etching times using a 3D stereo microscope, white light interferometer and profilometer. To quantitatively characterize the effectiveness of HF-based etching on eliminating the subsurface impurities, an energy dispersive X-ray spectrometer (EDS) is employed to test and compare the amounts of impurity defects before and after the acid etching process.

*2.3. Laser Damage Test*

The laser-induced damage thresholds (LIDTs) of fused silica optics before and after HF acid etching are measured following R-on-1 test protocol to validate the effect of impurity on laser damage resistance. In this test protocol, the laser pulse with fluence far below the damage threshold is initially applied. Then, incident lasers with pulse fluence increasingly ramped up are adopted until damage is observed. The LIDT is defined as the lowest fluence at which the damage occurs [44]. The laser damage experiment is carried out using a Q-switched Nd: YAG pump lasers (SGR-Exra-10, provided by the Beamtech Company (Beijing, China). It is capable of steadily providing 1064 nm and 532 nm wavelength lasers with a 1-Hz repetition rate and 10-ns pulse width. The detailed parameters of the laser damage test are listed in Table 3.

**Table 3.** The experimental parameters for laser damage test on fused silica.

| Pulse Width $\tau_p$/ns | Wavelength $\lambda$/nm | Repetition Rate $v$/Hz | Beam Diameter $D$/µm | Incident Angle $\theta_i$/deg | Laser Modal | Divergence Angle $\theta_d$/mrad |
|---|---|---|---|---|---|---|
| 10 | 355 | 1 | 390 | 0 | TEM$_{00}$ | $\leq$2.5 |

The setup used to test the LIDT of fused silica is sketched in Figure 4. The pump laser is effectively delivered and focused on the surface of tested fused silica sample by propagating through focusing lens, wedged splitter and reflection mirrors. The attenuator consisting of polarizer and half-wave plate is used to adjust the energy output of the laser system. The pulse energy of each laser shot is calculated and recorded by monitoring the partial light energy, split from the main pump lasers using a wedged splitter. The sample is firmly mounted on the 3D translation stage to assure its position is accurately adjusted during the laser damage test. The judgment of the laser damage initiation is crucial to the laser damage test in that the measured LIDT is directly determined by the laser energy at which the laser damage is initiated. The Normaski microscope camera with magnification ranging from 100 to 200 times is employed to image any permanent changes (i.e., laser-induced damage) on the tested sample surface. Furthermore, the laser damage morphology is also real-time monitored with a He-Ne laser scattering system as shown in Figure 4b. By travelling through the focus lens, the He-Ne probe lasers scattered from the tested surface can be focused on the photoelectric detector. The detector linked with a phase-locked amplifier is capable of recognizing the weak scattering light signal. When the laser damage is initiated on the tested surface, the signal becomes stronger. As a result, the laser damage can be precisely judged by real-time monitoring the change of the scattering light signal.

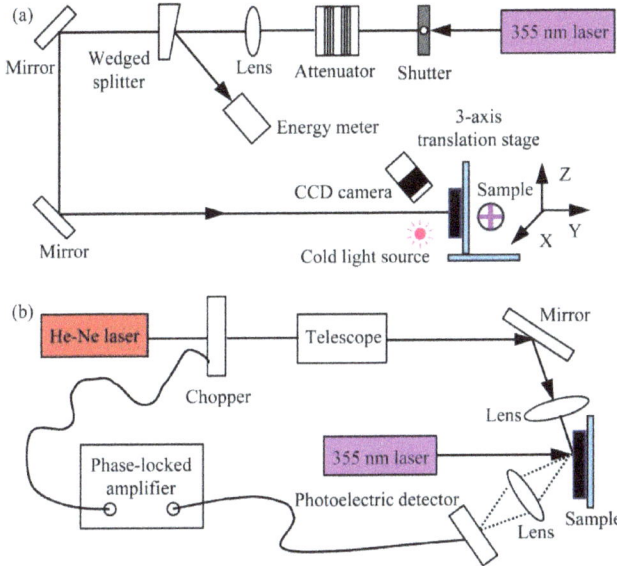

**Figure 4.** Schematic diagram of the laser path designed for the laser damage experiments (a) and the online monitoring system; (b) applied for detecting the laser damage initiation.

## 3. Results and Discussion

### 3.1. Temperature Distribution Caused by the Subsurface Impurity Defects

#### 3.1.1. Comparison of Temperature Distributions with and without Impurity Defects

Figure 5 shows the comparison of temperature distributions caused by fused silica optics with and without impurity defects under the irradiation of Gaussian laser pulse with a 355 nm wavelength, 10 ns pulse width and 10 J/cm$^2$ laser fluence. The typical subsurface impurity defects of $CeO_2$, Fe and Al are all considered with defect diameter of 1 μm that is located 1 μm (the distance between defect center and surface) beneath the optical surface. Figure 5a,b exhibits the temperature curves for fused silica without any impurity defect. It is shown that under the irradiation of Gaussian laser pulse, the temperature field on the silica surface is also distributed in a Gaussian profile. With the increase of laser loading time ($t < 10$ ns), the temperatures on the surface rise gradually, and the maximum temperature at the center of Gaussian profile ($r_p = 0$ mm) can reach the peak of 1650 K when the laser pulse lasts for 10 ns (pulse width). The peak temperature is lower than the melting point (2173 K) of fused silica so that no laser damage will occur under this condition. The evolution of peak temperature ($r_p = 0$ mm) in the heating ($t < 10$ ns) and cooling ($t > 10$ ns) phases is exhibited in Figure 5b. One can see that when the laser stops heating, the peak temperature will dramatically decrease. The temperature gradually falls down thereafter to a stable and low value at room temperature (~300 K), when the laser pulse ends for $10^4$ ns. It means that the simulated temperature field would reach a stable state finally. It should be noted that the central-point temperature in Figure 5b shows a short plateau from $t = 15$ ns to $t = 30$ ns and the potential explanation is listed below. After the laser ends ($t > 10$ ns), when the decreasing surface temperature is close to that in the bulk, the effects of convection and radiation of the surface would make the surface temperature keep decreasing. The surface temperature would be then lower than that of the bulk, and the heat energy could be transferred from the bulk to the surface. Owing to the competitive contributions of natural convection, surface radiation and heat transfer, the central-point temperature may remain roughly a short constant.

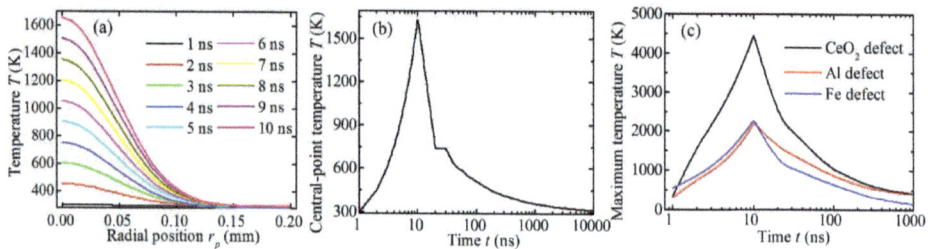

**Figure 5.** Comparison of temperature distributions caused by fused silica optics with and without impurity defects. (**a**) Temperature variation with respect to the radial position without impurity defect. Evolution of the peak temperature for fused silica optics without (**b**) and with (**c**) impurity defects of $CeO_2$, Al and Fe.

Figure 5c shows the variations of peak temperatures caused by impurity defects of $CeO_2$, Fe and Al. It can be seen that the peak temperature caused by $CeO_2$ defect rises more strongly than those caused by Fe and Al defects. The maximum temperature can reach 4458 K, which is much higher than the melting point of fused silica and the melting (2673 K) and boiling (3773 K) points of $CeO_2$. This means that the fused silica would be seriously damaged with the presence of a $CeO_2$ impurity defect. The maximum temperatures caused by Al and Fe impurity defects are both above 2200 K, which is also higher than the melting points of Al (933 K) and Fe (1808 K). Hence, the fused silica optics with Fe and Al impurity defects would also suffer from laser damage under the irradiation of an intense laser pulse. It can be concluded that with the presence of impurity defects, the optical materials would strongly absorb the laser energy, melt or even boil, and finally incur optical breakdown under the irradiation of a 355 nm wavelength, 10 ns pulse width and 10 J/cm² laser fluence.

The evolutions of temperature distributions caused by impurity defects are presented in Figure 6. As shown in Figure 6a–c, due to the strong absorption of $CeO_2$ defect, the temperature on the upper surface of impurity defect greatly increases in the heating process of laser pulse ($t < 10$ ns). When the laser pulse ends, the absorbed thermal energy is gradually diffused and transferred upward to the fused silica surface and downward to the bulk of the spherical $CeO_2$ defect. This means that the temperature distribution around the defect is non-symmetrical along the z-direction and should be highly dependent on their distances relative to the location of the initial incident lasers. Figure 6d,e shows the temperature distribution along the central axis for $CeO_2$ and Al defects at various times. One can see that, as the laser pulse acts, the temperature inside Al impurity defect (z-position ranging from 8.5 μm to 9.5 μm) becomes more and more uniform. The temperature is totally uniformly distributed in the bulk of Al defect when $t > 30$ ns (the curve keeps horizontally stable). However, for the case of the $CeO_2$ impurity defect, its induced temperature distribution inside impurity defect keeps barely uniform. The fact that the heat conduction capacity of Al is much better than that of $CeO_2$ should be responsible for this phenomenon.

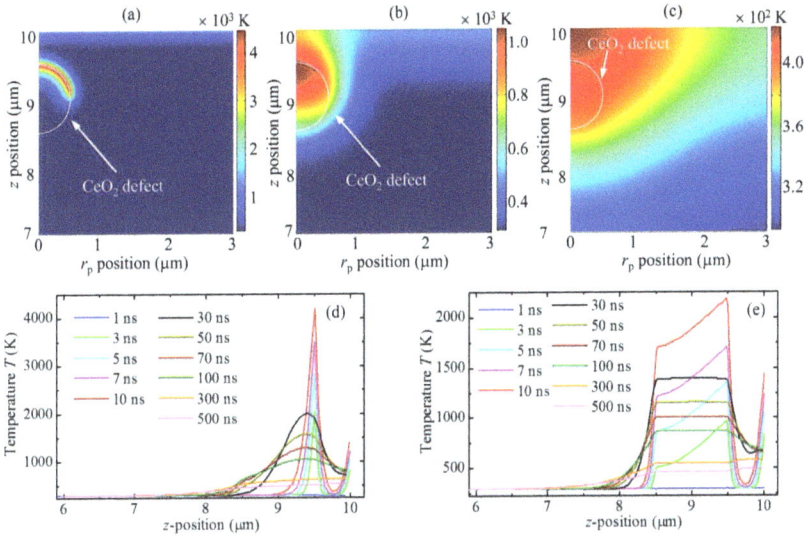

**Figure 6.** The temperature distributions on the longitudinal cross-section ($z$-$r_p$ plane) caused by $CeO_2$ defect for laser loading times of (**a**) 10 ns; (**b**) 100 ns; (**c**) 1000 ns and (**d**) the temperature distributions along the central axis ($r_p = 0$ μm) caused by $CeO_2$ and Al (**e**) defects for various laser loading times.

### 3.1.2. Effect of Impurity Defect Parameters on Temperature Distribution

In order to figure out the most dangerous impurity defect to the laser damage resistance of fused silica, we investigate the temperature distributions caused by impurity defects with various structural parameters (e.g., defect radius $r$, defect depth $d$). The effect of defect structural parameters on its induced temperature rise is demonstrated in Figure 7. As shown in Figure 7a, after being exposed in lasers for 10 ns, the maximum temperatures caused by impurity defects ascend sharply and then decrease gradually with the increase of defect radius. The critical defect radius corresponding to the highest temperatures for Al and Fe impurity defects is 0.15 μm. The highest temperatures are 3037 K and 2965 K, respectively, for the Al and Fe defects. Meanwhile, for the $CeO_2$ defect, the critical radius is 0.15 μm with the highest temperature of 4531 K. Temperature with this order of magnitude is high enough to incur matter evaporation, or even plasma formation, which would consequently induce local volume expansion, surrounding material fracture and finally laser damage. The phenomenon of maximum temperature evolution with respect to the defect radius can be explained as follows: as the defect radius increases, the defect area irradiated by incident lasers will enlarge, resulting in more laser energy to be absorbed by the defect. However, the enlarged irradiation area of impurity defect will increase the energy loss by thermal diffusion as well. Besides, with the increase of defect radius, the increase rate of absorbed energy would gradually decrease owing to the Gaussian distribution of the incident laser energy. Based on this, there must be a critical defect radius existing between the heat absorbing and diffusing processes, at which the temperature would reach the summit. This is consistent with the results shown in Figure 7a. Figure 7b shows the dynamic evolution of maximum temperature versus the defect radius. One can see that during the laser loading period ($t < 10$ ns), the maximum temperature increases firstly and then decreases gradually similar to the results as shown in Figure 7a. When the laser pulse ends ($t > 10$ ns), the temperature keeps rising all the time with the increase of defect radius. This is because under the same heat diffusing circumstance, an impurity defect with large size will hold more thermal energy and correspondingly induce a higher temperature. This agrees with the explanation discussed above. The simulation results at the loading times before ($t = 1$ ns, 3 ns, 5 ns, 7 ns, 10 ns) and after ($t = 30$ ns, 50 ns, 70 ns, 100 ns, 300 ns) the incident laser heating process as shown in Figure 7b validate that the maximum temperatures caused by impurity defects with various radii generally take place at the ending time ($t = 10$ ns) of the laser pulse. Hence, the variations of maximum temperatures at $t = 10$ ns with respect to the defect radius and depth are summarized in Figure 7a,c to obtain the most dangerous impurity defects in the polished fused silica optics.

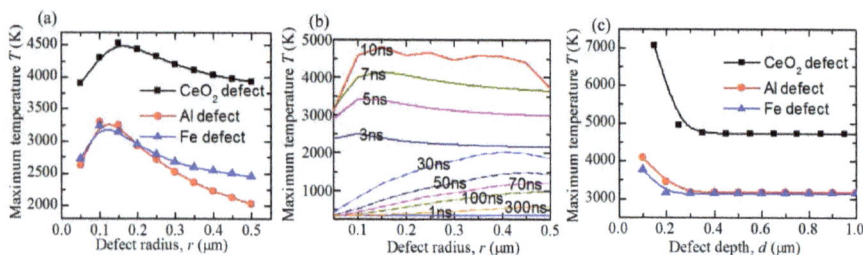

**Figure 7.** Evolutions of the maximum temperatures caused by various impurity defects with respect to the defect radius (**a**) and depth (**c**) during the laser loading period of 10 ns. (**b**) The dependence of maximum temperature on defect radius for CeO$_2$ defects under various laser loading times.

Figure 7c exhibits the dependence of maximum temperature on the defect depth. It is shown that the maximum temperature caused by impurity defect is higher when it is located more closely to the fused silica surface. When the defect depth is deeper than 0.3 μm, the maximum temperature will keep constant as the defect depth increases. This is because the heat conductivity coefficient of ambient air (0.023 W·m$^{-1}$·K$^{-1}$) is lower than that of fused silica materials. For defect located more closely to the surface, the thermal energy is more prone to transfer among the ambient air. Under this circumstance, less energy will be lost by thermal diffusion. Hence, the temperature is higher when the defect resides closer to the silica surface. However, when the defect is located deeper than 0.3 μm beneath the surface, the effect of ambient air on thermal diffusion becomes negligible. From the simulations of temperature distributions shown above, we can conclude that impurity defects (especially for CeO$_2$) located closer than 0.3 μm away from the surface with defect radius ranging from 0.1 μm to 0.15 μm are most dangerous defects for lowering the laser damage resistance of polished fused silica optics.

## 3.2. Thermal Stress Distribution Caused by the Subsurface Impurity Defects

### 3.2.1. Comparison of Thermal Stress Distributions with and without Impurity Defect

By numerically solving Equations (1) and (4) with finite element method (FEM), the physics of heat conduction and solid mechanics involved in the intense laser irradiation process can be well coupled for obtaining the thermal stress distribution caused by impurity defects. The thermal stress could be applied to describe the negative effect of impurity defects on the mechanical property and wavefront distortion of fused silica optics with high precision surfaces. In this section, the von Mises equivalent stress is employed to evaluating the yielding behavior of fused silica material with the presence of subsurface impurity defects. Figure 8 presents the comparison of von Mises equivalent stress on fused silica with and without impurity defects. The parameters of incident lasers are the same to those applied in Section 3.1 (10 J/cm$^2$, 355 nm, 10 ns) and the defects of CeO$_2$, Fe and Al are all investigated with a 1 μm defect radius and a 1 μm defect depth. The negative effect of impurity defects on laser damage resistance of fused silica optics is primarily related to the maximum temperature and stress during the laser irradiating process. Hence, in this work, we only concentrate on the transient simulation results of the maximum temperature and stress, which generally take place at roughly 10 ns when the laser pulse ends.

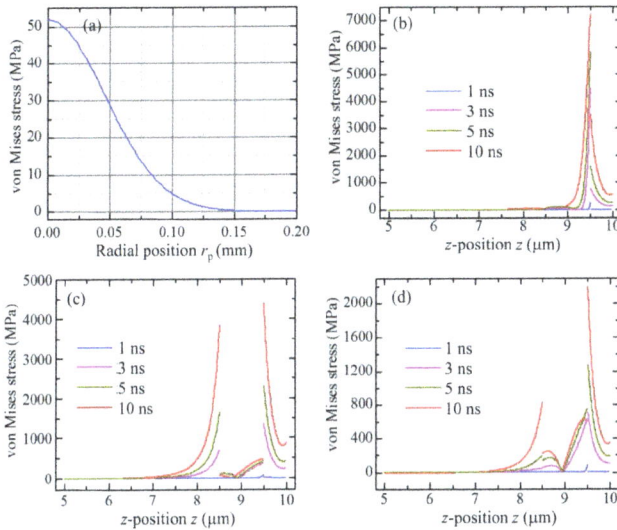

**Figure 8.** Comparison of the von Mises thermal stress distributions with and without impurity defects. (a) Thermal stress distribution along the radial direction ($z = 10~\mu m$) on fused silica surface without impurity defect. The maximum stress at $r_p = 0~\mu m$ could be applied in the comparison with those caused by the impurity defects. The von Mises thermal stress distribution along the central axis ($r_p = 0~\mu m$) caused by (b) $CeO_2$; (c) Al and (d) Fe defects for various laser loading times. The $z$-positions of 8.5 $\mu m$ and 9.5 $\mu m$ are the intersection points of fused silica with the downmost and upmost parts of the impurity defect.

As shown in Figure 8a, for fused silica optics without a subsurface defect, the induced von Mises stress appears in standard Gaussian distribution under the laser irradiation. The maximum thermal stress is 52 MPa, which is far lower than the strength of fused silica (800–1000 MPa). It means that the mechanical property and surface deformation of fused silica would be not strongly affected under the laser irradiation with this level of power. For the thermal stress in fused silica optics with impurity defects, it is demonstrated in Figure 8b–d that different types of impurity defects can induce different thermal stress distribution under the intense laser irradiation. The highest thermal stress is induced by the $CeO_2$ impurity defect, followed by Al and Fe defects. The maximum von Mises equivalent thermal stress induced by $CeO_2$ defect is 7432 MPa located on the topmost surface of defect and the maximum stress on the fused silica material is 3617 MPa. For Al and Fe defects, the induced maximum Mises equivalent thermal stresses are 4476 MPa and 2236 MPa, respectively. By comparing the simulated thermal stress distributions and thermodynamic parameters of Al and Fe defects, one can see that the coefficient of thermal expansion plays a dominant role in producing the higher thermal stress. It is also shown in Figure 8c,d that sharp changes exist at the intersections of fused silica material and impurity defect in the evolutions of von Mises stress caused by Al and Fe defects (the thermal stress curves have discontinuous points at $z$-positions of 8.5 $\mu m$ and 9.5 $\mu m$). This phenomenon can be explained below: (1) under the intense laser irradiation, the temperature rise of the impurity defect is greater than that of surrounding fused silica material and the local concentrated thermal load may be formed; (2) the thermal expansion coefficient of fused silica is much smaller than that of impurity defects that the impurity defect undergoing thermal expansion would be spatially constrained by the local surrounding silica materials; and (3) a big difference exists in the rigidity property of neighboring FEM elements due to the natural mechanical properties of the involved materials. However, for $CeO_2$ defect as shown in Figure 8b, since the thermal energy has been not already transferred to the downmost part ($z$-position of 8.5 $\mu m$) of impurity defect, the temperature rise in this region is very small. Besides, the $CeO_2$ defect and fused silica materials share a similar thermal expansion coefficient. Thus, for $CeO_2$ defect, the thermal stress curve at the lower intersection point of fused silica and impurity defect (z-position of 8.5 μm) behaves continuously. While for the upper intersection point (z-position of 9.5 μm), the temperature of impurity defect is much higher than that of fused silica. Under this situation, an obvious heat load would be probably formed, resulting in evident thermal stress discontinuity as shown in Figure 8b. It can be concluded that the impurity defect would lower the laser damage resistance of fused silica optics by its induced thermal expansion and stress. The thermal stress of impurity defect may incur high stress in the surrounding silica materials, which would exceed the compressive strength of fused silica optics and finally lead to its fracture and breakdown.

### 3.2.2. Effect of Impurity Defect Parameters on Its Induced Thermal Stress

In order to discern the most dangerous impurity defect to the laser damage resistance of fused silica optics, we also investigate the von Mises thermal stress caused by impurity defects with various structural parameters (e.g., defect radius $r$, defect depth $d$). Since the $CeO_2$ impurity defect would induce the highest temperature and thermal stress as discussed above, here we only consider the effect of $CeO_2$ defect parameters on its induced thermal stress, which is shown in Figure 9. As shown in Figure 9a, with the increase of defect radius, the von Mises stress ascends firstly and then decreases gradually. The maximum thermal stress reaches the summit (7420 MPa) when the defect radius is 0.15 μm (critical defect radius). One can see from Figure 9b that impurity defect located more closely to the silica surface would induce higher maximum thermal stress. The peak maximum stress of fused silica is 8739 MPa, and when the defect depth is larger than 0.3 μm, it remains roughly stable. The changing tendency of maximum thermal stress with respect to defect parameters is similar to that of maximum temperature shown in Figure 7. It is worth noting that the thermal stress of fused silica is higher than that of $CeO_2$ defect as shown in Figure 9b. This is because the defect radius is so small ($r$ keeps constant at the critical radius of 0.15 μm) that the heat energy can be promptly transferred to the surrounding fused silica materials. As a result, fused silica with higher heat capacity would induce higher thermal stress.

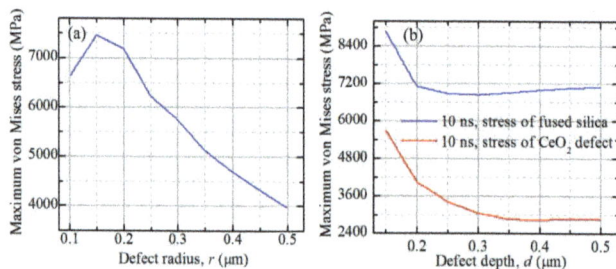

**Figure 9.** Evolution of the maximum von Mises stress caused by $CeO_2$ impurity defect with respect to the defect radius (**a**) and depth (**b**) when the fused silica surface is irradiated by incident lasers for 10 ns. The maximum thermal stresses in fused silica and $CeO_2$ defect are both presented in (**b**).

From the simulation results of defect-induced temperature and thermal stress distributions discussed above, one can see that the transient local high temperature can be produced by heat absorbing of impurity defects under the irradiation of intense lasers. The extremely high temperature would trigger the plasma formation inside the impurity defect, result in material expansion, and finally initiate the fracture and breakdown of the surrounding silica materials. In addition, the huge thermal stress caused by the impurity defect far exceeds the intrinsic strength limit of the material, making the fused silica optics susceptible to be crushed down. Furthermore, in the actual process of intense laser irradiation, the high temperature and thermal stress caused by the impurity defect could act together, leading to even worse damage of fused silica optics. Based on the results above, we can conclude that the temperature and thermal stress caused by impurity defects strongly depends on the defect species and structural parameters. The $CeO_2$ defect can induce higher temperature and thermal stress than Al and Fe defects, especially when it is located more closely to the silica surface. The most dangerous subsurface impurity defect is the $CeO_2$ defect with a diameter of 0.3 μm, located less than 0.15 μm beneath the surface, which would induce local temperature up to 7073 K and von Mises equivalent thermal stress up to 8739 MPa. Therefore, we should pay close attention to control and remove this dangerous kind of subsurface defects in the practical manufacturing process of fused silica optics with high precision surfaces. By considering the synthetic effects of temperature and thermal stress caused by impurity defects, it can be concluded from the simulation results shown in Figures 8b–d and 7 that the $CeO_2$ defect plays the dominant role in decreasing the laser damage resistance of fused silica, followed by Al and Fe defects. It has been also experimentally observed by Neauport et al. [17,25] that the laser damage density of silica optics is most closely correlated to the cerium content, followed by Al and Fe defects. The correlation coefficients of damage density to the Ce, Al and Fe impurity defects are 0.99, 0.16 and 0.001, respectively. Hence, the simulation results of temperature and stress in this work are qualitatively supported by the previously experimental results and the present results would offer potential theoretical explanations on the experimental observations of the effect of impurity defects on the laser damage resistance of fused silica optics.

### 3.3. Removal of Subsurface Impurity Defects by HF Acid Etching

The simulation results of temperature and thermal stress caused by subsurface impurity defects presented in Sections 3.1 and 3.2 indicate that the impurity defects included in the SSD lasers play a crucial role in lowering the laser damage resistance of polished fused silica. In order to improve the laser damage resistance of silica optics, HF acid etching assisted with buffered oxide etch (BOE) is applied to remove the SSD layers. Then, the etching rate, surface quality and impurity contents are all investigated, and the laser damage threshold is tested to validate the effect of impurity defects on the laser damage resistance of fused silica optics.

#### 3.3.1. Etching Rate of Fused Silica

Figure 10 presents the measured variation of etching rate as a function of etching time and etching depth using a stylus profilometer on the basis of a test scheme shown in Figure 3. One can see that in the initial etching process, the etching rate exhibits a sharp decrease from 46.45 nm/min to 28.81 nm/min. As the etching proceeds, the etching rate gradually descends to 23 nm/min and remains roughly stable. This phenomenon can be ascribed to the different speeds of chemical reaction of HF acid with body material ($SiO_2$) and hydrolysis product in form of hydrated silica gel ($\equiv$Si–OH). The product of hydrated silica gel is formed following the process below:

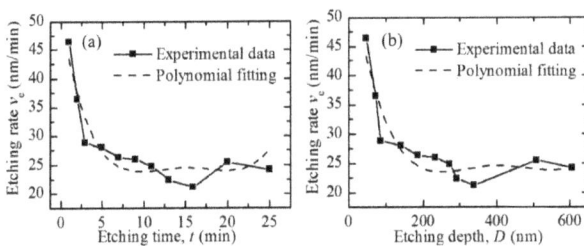

$$\equiv Si\text{--}O\text{--}Si \equiv \; + H_2O \; \rightarrow \; 2 \equiv Si\text{--}OH \tag{5}$$

**Figure 10.** The evolution of etching rate with respect to the etching time (**a**) and etching depth (**b**).

Besides, the re-deposition layer produced in the grinding and polishing processes has a loose structure, which would enlarge its contact area with BOE and acid solvents. Hence, the etching rate appears very large in the initial etching phase. However, the curve shows a downtrend behavior because of the decreasing concentration of hydrated silica gel as the etching proceeds. When the re-deposition layer is totally removed, and the pure substrate of fused silica is exposed to the HF-based solvent, the etching rate would change slightly and keep almost constant as shown in Figure 10. Owing to the random motions of grinding and polishing particles, the density and depth of SSD defects on fused silica optics are somewhat different, even though the same processing parameters are applied. For this reason, the etching rate fluctuates with 10% amplitude in the steady etching process as shown in Figure 10. By polynomial fitting the etching rate data, we can conclude that a 200-nm etching depth and an 8-min etching time are required for steadily etching the SSD layers of polished silica optics. Based on the aforementioned discussions, it can be also inferred that the thickness of ground and polished silica optics is roughly 200 nm, which is consistent with most of the previously reported experimental results [24,26,45].

### 3.3.2. Surface Quality and Impurity Content of Fused Silica Etched by HF Acid

As the HF acid etching proceeds, not only the impurity defects included in the re-deposition layers are removed, but the subsurface cracks and scratches are blunted [26,27]. The optical quality of etched fused silica surfaces would have great changes and the surface roughness value is an important index in evaluating the quality of optical surface. Thus, the evolution of surface roughness value as the HF acid etching proceeds is investigated using the white light interferometer and profilometer to check the effect of acid etching on the surface quality. Figure 11 shows the tested evolution of surface roughness with respect to the etching time (or etching depth) for fused silica etched with HF-based solvent (5% HF and 10% NH$_4$F). It is shown that the surface roughness value Ra experiences three changing phases as the HF etching proceeds. During phase 1 with the etching depth smaller than 500 nm, the Ra ascends from 2.6 nm to 4.2 nm. The decrease of surface quality (rise of roughness) arises from the appearance of subsurface plastic scratches and cracks as the topmost re-deposition layers are removed. During phase 2 with the etching depth ranging from 0.5 μm to 2.15 μm, the Ra value descends from 4.2 nm to 3.3 nm due to the passivation of subsurface damage under the assistance of acid etchants. The subsurface cracks and scratches are etched and blunted, resulting in the improvement of surface quality. The surface roughness increases again during phase 3 for etching depth larger than 2.15 μm. The reproduction of scratches and opening of cracks with the increase of etching depth should be responsible for this results.

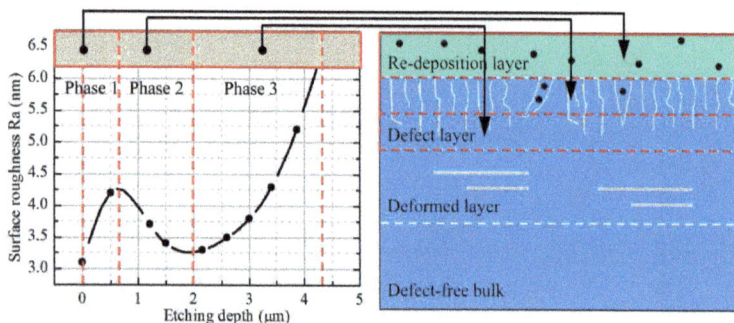

**Figure 11.** The evolution of surface roughness of etched fused silica optics as the acid etching proceeds and its corresponding etching depths.

Figure 12 presents the evolution of surface morphology of etched fused silica with various etching periods, which are tested by a 3D stereo microscope and white light interferometer. It can be seen that with the increase of etching period, the subsurface defects (e.g., scratches, cracks, pits and so on) emerge and the defect population become larger. This is the further evidence for the great decrease of surface quality during phase 3 as shown in Figure 11. It should be noted that with a mixed etchant of 5% HF and 10% NH₄F, as applied in Figure 12, the re-deposition layer is totally removed when the etching period is 3 min. Thus, all of the tested surface morphology in Figure 12 corresponds to phase 3 shown in Figure 11.

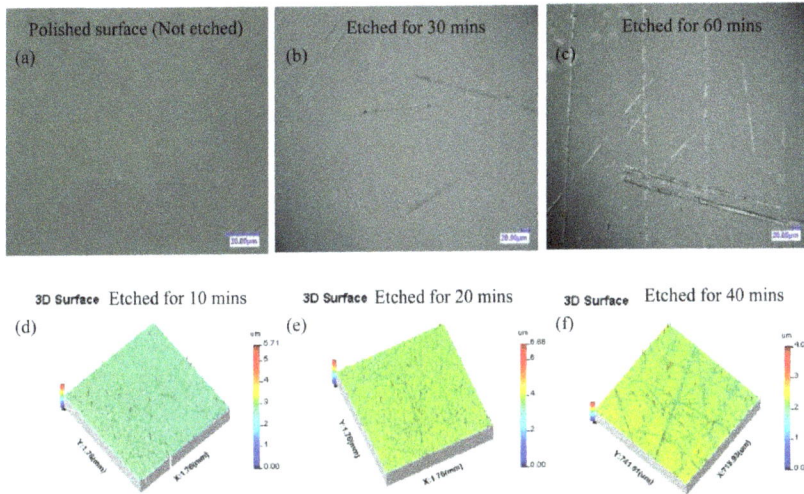

**Figure 12.** The evolution of surface morphology of etched fused silica under various etching periods. The etched surfaces are tested with a 3D stereo microscope for etching periods of (**a**) 0 min; (**b**) 30 min and (**c**) 60 min The surface morphologies of etched surfaces are also tested with white light interferometer for etching periods of (**d**) 10 min; (**e**) 20 min and (**f**) 40 min.

To quantitatively evaluate the effectiveness of HF-based acid etching in removing the subsurface impurity defects, the comparison of defect contents on fused silica surfaces before and after acid etching is investigated using energy dispersive X-ray spectrometer (EDS). Figure 13 shows the comparison of defect content and the evolution of Ce content with the increase of etching time. For the defect content before acid etching as shown in Figure 13a, one can see that the contents of Si and O are two types of elements most distributed on the polished silica surface. They originate from the body material of fused silica (SiO₂) and satisfy their proportional relationship of mass fraction. Besides Si and O, other impurity defects (e.g., Ce, Fe, Cu and Se) are also observed with the maximum defect content of 7.45 wt % for Ce impurity. The results obtained in Section 3.1 and 3.2 indicate that the CeO₂ impurity defect plays the dominant role in lowering the laser damage resistance of polished fused silica optics. Hence, we take Ce defect as an example to investigate its elimination when the SSD layers of silica optics are etched with HF etchant. The tested impurity contents on etched silica surfaces are presented in Figure 13b. It is shown that the contents of most of the impurity defects are effectively lowered by acid etching, especially for the Ce impurity defect, whose content largely descends from 7.45 wt % to 3.24 wt %. It means that 56.5% of the CeO₂ impurity defects could be removed by HF acid etching and the heat absorption caused by impurity defects would be greatly mitigated to improve the laser damage resistance of fused silica. The evolution of Ce defect content with respect to the etching time is exhibited in Figure 13c. It can be seen that the content of Ce defect decreases gradually and becomes roughly stable when the fused silica optics are etched for a sufficient time. This phenomenon can be explained as follows. In the initial etching stage, the defect content decreases dramatically due to the removal of re-deposition layers, among which a majority of impurity defects are located and distributed. After etching for 10 min, when the re-deposition layers are totally removed, the subsurface scratches and cracks appear and some of the impurity defects hidden inside or adsorbed to the cracks could be removed and washed away. The defect content therefore shows gradual decrease for etching time from 10 min to 50 min. When the silica optics are etched for longer than 50 min, the defect content keeps almost constant since the impurity defects adsorbed to the cracks are then difficult to be removed by ordinary etchant washing. From the discussions above, it can be concluded that the impurity defects can be effectively removed by HF-based acid etching. The acid etching shows the best effectiveness in removing the impurity defects, when the re-deposition layers are being etched. The mechanical properties and laser damage resistance of fused silica would be greatly improved by controlling the heat absorption caused by these impurity defects.

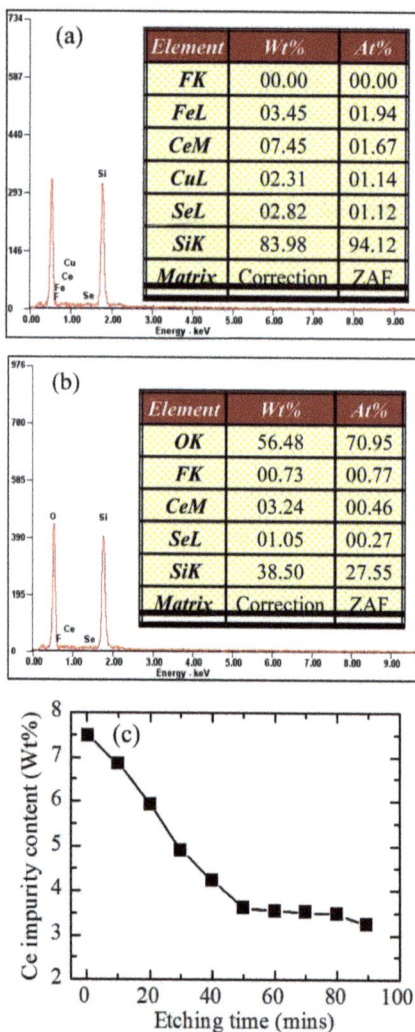

**Figure 13.** Tested results of impurity defects on polished silica surfaces before and after acid etching: (**a**) impurity contents before HF acid etching; (**b**) impurity contents after HF acid etching for 90 min; (**c**) Variation of Ce defect content for various etching times.

### 3.3.3. Laser Damage Resistance of Etched Fused Silica Surface

Now that the contents of impurity defects are effectively controlled, another key index, laser-induced damage threshold (LIDT) of etched fused silica optics should be checked as well. The LIDTs of fused silica etched for various times using etchant of 5% HF and 10% $NH_4F$ are measured with the testing protocol and setup described in Section 2.3. A 10-min step of etching period is chosen in the HF-etching procedure and the reason as follows. Firstly, it can be derived from Figure 10 that an 8-min etching period and a 200 nm etching depth are required to ensure a steady etching rate for the polished fused silica optics with SSD layers. It means that the impurity defects mainly distributed among the uppermost re-deposition layer could be effectively removed within 10 min. Besides, the general trend of the tested Ce impurity content shown in Figure 13c would be clearly recognized with an etching period of 10 min.

The tested LIDTs for various etching times are shown in Figure 14a. It can be seen that the LIDT rises firstly (etching for less than 10 min) and then gradually decreases (etching for more than 10 min) as the etching proceeds. It is indicated by the results in Figure 10 that the re-deposition layers containing most of the impurity defects and hydrolysis products are initially removed during the first 8-min etching times. Combined with the results shown in Figure 14a, one can conclude that not only the impurity defects are effectively removed, but the LIDT of etched fused silica optics could be enhanced from 24.4 $J/cm^2$ to 27.1 $J/cm^2$ (355 nm, 10 ns). However, when the re-deposition layers are fully removed, the LIDT presents gradual decrease as the etching proceeds due to the appearance of subsurface scratches and cracks. Based on this, in the practical etching process of fused silica optics, the etching time should be accurately and strictly controlled to prevent the appearing and deepening of subsurface cracks and scratches. The morphological comparisons of laser damage spots on un-etched and etched fused silica optics are exhibited in Figure 14b,c. It can be seen that the damage spot on an un-etched silica surface is very large, consisting of many local small and concentrated damage sites. Besides, the damaged silica materials are all peeled off. As for the damage spots on etched silica surfaces, they are very small and discretely distributed right on the surface scratches. The difference in laser damage morphology on etched and un-etched silica surfaces indicates that the un-etched surfaces possess more impurity defects, which are the main initiators for inducing serious optical damage. When the polished fused silica surfaces are etched by HF acid, the amounts of impurity defects are greatly reduced, and correspondingly the laser damage resistance of silica optics could be effectively improved. This is consistent with the tested results of impurity defect contents on etched and un-etched fused silica surfaces. The laser damage experiment directly validates the negative effect of impurity defects on the laser damage resistance of fused silica. Since the $CeO_2$, Al and Fe impurity defects are very difficult to be separated from each other and a silica sample with a single impurity defect is difficult to prepare in actual experiments, the respective effects of single impurity defects on laser damage resistance of fused silica optics are hard to be directly and quantitatively validated. Furthermore, the transient temperature caused by absorptive impurity defects is very high and it only lasts for such a short time that experimentally acquiring the transient high temperature is quite a challenge. Thus, it is very difficult to detect the accurate temperature evolution on the laser-irradiated silica surface to directly justify the simulation results in this work. However, the experimental results of the laser damage test show that the impurity defects are probably the absorbing initiators that incur the laser damage event. This indirectly validates the simulation results that the much higher temperature and stress caused by impurity defects in comparison to defect-free surfaces are the potential underlying mechanisms for explaining the low laser damage threshold of polished fused silica optics.

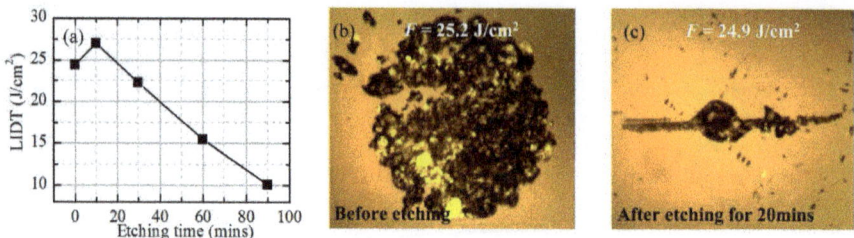

**Figure 14.** Comparison of the laser damage performance for fused silica optics before and after HF etching. (**a**) Variation of laser-induced damage thresholds (LIDTs) of etched silica surface under various etching times. The typical laser damage morphologies on un-etched (**b**) and etched (**c**) silica surfaces. The microscopic images in (**b**,**c**) are both magnified by 40 times.

The decrease of LIDT as shown in Figure 14a and the initiation of laser damage shown in Figure 14c should be blamed for the appearance of subsurface cracks and scratches. To solve this issue, two other parallel processing techniques, $CO_2$ laser conditioning [8] and thermal annealing [9,10], have been also being developed to close and heal the cracks and scratches by heat softening for mitigating the negative effect of subsurface defects, which is beyond the scope of the present work. It is worth noting that though the amount of $CeO_2$ defects as shown in Figure 13 can be reduced by more than a half, some residual impurity defects are still adsorbed on the etched surfaces or hidden beneath the subsurface cracks. The residual impurity defects would limit the capacity of the HF-based acid etching technique in promoting the laser damage resistance of fused silica optics. Therefore, based on HF-based etching, some new techniques (e.g., assistance of megasonic agitation and addition of chelating agent adoption) are being explored and developed to further remove the subsurface impurity defects of fused silica optics. In the experiments, the variation of the impurity defect population with respect to the increase of etching time is indirectly derived by characterizing the etching rate, impurity element content, laser damage threshold, surface roughness and morphology of etched silica surfaces. The errors in the etching process would bring in potential deviations of these characterized features to the actual case. Hence, the simulation results of the negative effects of impurity defects have been not quantitatively and accurately validated, and new techniques of testing the transient high temperature and analyzing the exact impurity defect content should be developed in the future to further justify the role of impurity defects in decreasing the laser damage resistance of the polished fused silica optics.

## 4. Conclusions

To evaluate the effects of impurity defects on the laser damage resistance of fused silica, the distributions of temperature and thermal stress caused by absorptive defects are investigated by numerically solving the heat conduction and thermo-elastic equations involved in the laser irradiation process. The simulation results indicate that the presence of impurity defects would induce a dramatic rise in temperature and thermal stress. The maximum temperature and thermal stress in fused silica can exceed its melting point and compressive strength, resulting in optical breakdown of optical materials. The effect of impurity defects on the laser damage resistance is dependent on the species, size and spatial position of defects. The $CeO_2$ defect plays the dominant role in lowering the laser damage resistance, followed by Fe and Al defects. The defects located more closely to the silica surfaces would incur higher local temperature. It is further concluded that the $CeO_2$ defects with radius of roughly 0.3 μm, which reside 0.15 μm beneath the silica surfaces, are the most dangerous defects affecting the laser damage resistance of fused silica optics.

With the HF-based etching method, the impurity defects on fused silica are removed and the surface quality, defect content and laser damage threshold of etched surfaces are experimentally tested to evaluate the etching effectiveness. The results show that the negative effect of impurity defects on the laser damage resistance can be effectively mitigated by HF acid etching. As the acid etching proceeds, the re-deposition and subsurface defect layers are successively removed, resulting in the evolution of surface roughness in three phases. The EDS and laser damage experiments validate that the number of dangerous $CeO_2$ defects on fused silica can be decreased by more than a half and the LIDT can be improved to 27.1 $J/cm^2$ (355 nm, 10 ns) by removing the re-deposition layers. It is proved that the HF-based acid etching is able to effectively control the bad effect of impurity defects on the laser damage resistance and the further improvement of LIDT is limited by the deep scratches and cracks. This work can not only contribute to the understanding of laser-induced damage mechanisms on large-aperture ultraviolet optical components, but also provide theoretical foundations for the post-processing of ultra-precision machined fused silica optics.

**Acknowledgments:** The authors gratefully acknowledge the financial support from the Science Challenge Project (Grant No. JCKY2016212A506-0503) and the Key Laboratory Fund of Ultra-precision Machining Technology at the China Academy of Engineering Physics (Grant No. KF14007).

**Author Contributions:** Jian Cheng and Jinghe Wang conceived the work and supervised the preparation of the manuscript. Jian Cheng, Lei Zhang and Jing Hou performed the FEM simulations and analyzed the data. Jian Cheng, Jinghe Wang and Hongxiang Wang designed the experiments of acid etching and laser damage tests. Jian Cheng, Jing Hou, Lei Zhang and Hongxiang Wang fabricated and characterized the sample surfaces. Jian Cheng wrote the manuscript with the input from other coauthors. All authors reviewed the manuscript.

**Conflicts of Interest:** The authors declare no conflict of interest.

## References

1.  Stolz, C.J. The National Ignition Facility: The path to a carbon-free energy future. *Philos. Trans. R. Soc. A* **2012**, *370*, 4115–4129. [CrossRef] [PubMed]
2.  Hurricane, O.A.; Callahan, D.A.; Casey, D.T.; Celliers, P.M.; Cerjan, C.; Dewald, E.L.; Dittrich, T.R.; Döppner, T.; Hinkel, D.E.; Berzak Hopkins, L.F.; et al. Fuel gain exceeding unity in an inertially confined fusion implosion. *Nature* **2014**, *506*, 343–348. [CrossRef] [PubMed]
3.  Casner, A.; Caillaud, T.; Darbon, S.; Duval, A.; Thfouin, I.; Jadaud, J.P.; LeBreton, J.P.; Reverdin, C.; Rosse, B.; Rosch, R.; et al. LMJ/PETAL laser facility: Overview and opportunities for laboratory astrophysics. *High Energy Density Phys.* **2015**, *17*, 2–11. [CrossRef]
4.  Dunne, M. A high-power laser fusion facility for Europe. *Nat. Phys.* **2006**, *2*, 2–5. [CrossRef]
5.  Zheng, W.; Wei, X.; Zhu, Q.; Jing, F.; Hu, D.; Su, J.; Zheng, K.; Yuan, X.; Zhou, H.; Dai, W.; et al. Laser performance of the SG-III laser facility. *High Power Laser Sci. Eng.* **2016**, *4*, e21. [CrossRef]
6.  Campbell, J.H.; Hawley-Fedder, R.A.; Stolz, C.J.; Menapace, J.A.; Borden, M.R.; Whitman, P.K.; Yu, J.; Runkel, M.; Riley, M.O.; Feit, M.D.; et al. NIF optical materials and fabrication technologies: An overview. *Proc. SPIE* **2004**, *5341*, 84–101.
7.  Baisden, P.A.; Atherton, L.J.; Hawley, R.A.; Land, T.A.; Menapace, J.A.; Miller, P.E.; Runkel, M.J.; Spaeth, M.L.; Stolz, C.J.; Suratwala, T.I.; et al. Large optics for the National Ignition Facility. *Fusion Sci. Technol.* **2016**, *69*, 295–351. [CrossRef]
8.  Lamaignère, L.; Bercegol, H.; Bouchut, P.; During, A.; Néauport, J.; Piombini, H.; Razé, G. Enhanced optical damage resistance of fused silica surfaces using UV laser conditioning and $CO_2$ laser treatment. *Proc. SPIE* **2004**, *5448*, 952–960.
9.  Raman, R.N.; Matthews, M.J.; Adams, J.J.; Demos, S.G. Monitoring annealing via $CO_2$ laser heating of defect populations on fused silica surfaces using photoluminescence microscopy. *Opt. Express* **2010**, *18*, 15207–15215. [CrossRef] [PubMed]
10. Raman, R.N.; Negres, R.A.; Matthews, M.J.; Carr, C.W. Effect of thermal anneal on growth behavior of laser-induced damage sites on the exit surface of fused silica. *Opt. Mater. Express* **2013**, *3*, 765–776. [CrossRef]
11. Smith, A.V.; Do, B.T. Picosecond-nanosecond bulk damage of fused silica at 1064 nm. *Proc. SPIE* **2008**, *7132*, 71321E.
12. Stäblein, J.; Pöhl, K.; Weisleder, A.; Gönna, G.; Töpfer, T.; Hein, J.; Siebold, M. Optical properties of $CaF_2$ and $Yb3^+$: $CaF_2$ for laser applications. *Proc. SPIE* **2011**, *8080*, 808002.
13. Azumi, M.; Nakahata, E. Study of relation between crystal structure and laser damage of Calcium Fluoride. *Proc. SPIE* **2010**, *7842*, 78421U.

14. Singh, N.B.; Suhre, D.R.; Balakrishna, V.; Marable, M.; Meyer, R. Far-infrared conversion materials: Gallium Selenide for far-infrared conversion application. *Prog. Cryst. Growth Charact. Mater.* **1998**, *37*, 47–102. [CrossRef]

15. DeYoreo, J.J.; Burnham, A.K.; Whitman, P.K. Developing $KH_2PO_4$ and $KD_2PO_4$ crystals for the world's most powerful laser. *Int. Mater. Rev.* **2002**, *47*, 113–152. [CrossRef]

16. **Negres, R.A.; Zaitseva, N.P.; DeMange, P.; Demos, S.G. Expedited laser damage profiling of $KD_xH_{2-x}PO_4$** with respect to crystal growth parameters. *Opt. Lett.* **2006**, *31*, 3110–3112. [CrossRef] [PubMed]

17. Neauport, J.; Lamaignere, L.; Bercegol, H.; Pilon, F.; Birolleau, J.C. Polishing-induced contamination of fused silica optics and laser induced damage density at 351 nm. *Opt. Express* **2005**, *13*, 10163–10171. [CrossRef] [PubMed]

18. Papernov, S.; Schmid, A.W. Laser-induced surface damage of optical materials: Absorption sources, initiation, growth, and mitigation. *Proc. SPIE* **2008**, *7132*, 71321J.

19. Wang, J.; Li, Y.; Han, J.; Xu, Q.; Guo, Y. Evaluating subsurface damage in optical glasses. *J. Eur. Opt. Soc. Rapid Publ.* **2011**, *6*, 11001.

20. Camp, D.W.; Kozlowski, M.R.; Sheehan, L.M.; Nichols, M.; Dovik, M.; Raether, R.; Thomas, I. Subsurface damage and polishing compound affect the 355-nm laser damage threshold of fused silica surfaces. *Proc. SPIE* **1998**, *3244*, 356–364.

21. Feit, M.D.; Rubenchik, A.M. Influence of subsurface cracks on laser induced surface damage. *Proc. SPIE* **2004**, *5273*, 264–272.

22. Neauport, J.; Cormont, P.; Legros, P.; Ambard, C.; Destribats, J. Imaging subsurface damage of grinded fused silica optics by confocal fluorescence microscopy. *Opt. Express* **2009**, *17*, 3543–3554. [CrossRef] [PubMed]

23. Liu, H.; Huang, J.; Wang, F.; Zhou, X.; Jiang, X.; Dong, W.; Zheng, W. Photoluminescence defects on subsurface layer of fused silica and its effects on laser damage performance. *Proc. SPIE* **2015**, *9255*, 92553V.

24. Hu, G.; Zhao, Y.; Liu, X.; Li, D.; Xiao, Q.; Yi, K.; Shao, J. Combining wet etching and real-time damage event imaging to reveal the most dangerous laser damage initiator in fused silica. *Opt. Lett.* **2013**, *38*, 2632–2635. [CrossRef] [PubMed]

25. Neauport, J.; Cormont, P.; Lamaignère, L.; Ambard, C.; Pilon, F.; Bercegol, H. Concerning the impact of polishing induced contamination of fused silica optics on the laser-induced damage density at 351 nm. *Opt. Commun.* **2008**, *281*, 3802–3805. [CrossRef]

26. Suratwala, T.I.; Miller, P.E.; Bude, J.D.; Steele, W.A.; Shen, N.; Monticelli, M.V.; Feit, M.D.; Laurence, T.A.; Norton, M.A.; Carr, C.W.; et al. HF-based etching processes for improving laser damage resistance of fused silica optical surfaces. *J. Am. Ceram. Soc.* **2011**, *94*, 416–428. [CrossRef]

27. Ye, H.; Li, Y.; Yuan, Z.; Wang, J.; Yang, W.; Xu, Q. Laser induced damage characteristics of fused silica optics treated by wet chemical processes. *Appl. Surf. Sci.* **2015**, *357*, 498–505. [CrossRef]

28. Ye, H.; Li, Y.; Zhang, Q.; Wang, W.; Yuan, Z.; Wang, J.; Xu, Q. Post-processing of fused silica and its effects on damage resistance to nanosecond pulsed UV lasers. *Appl. Opt.* **2016**, *55*, 3017–3025. [CrossRef] [PubMed]

29. Stevens-Kalceff, M.A.; Wong, J. Distribution of defects induced in fused silica by ultraviolet laser pulses before and after treatment with a $CO_2$ laser. *J. Appl. Phys.* **2005**, *97*, 113519. [CrossRef]

30. Hrubesh, L.W.; Norton, M.A.; Molander, W.A.; Donohue, E.E.; Maricle, S.M.; Penetrante, B.M.; Brusasco, R.M.; Grundler, W.; Butler, J.A.; Carr, J.W.; et al. Methods for mitigating surface damage growth on NIF final optics. *Proc. SPIE* **2002**, *4679*, 23–33.

31. Menapace, J.A.; Davis, P.J.; Steele, W.A.; Wong, L.L.; Suratwala, T.I.; Miller, P.E. MRF applications: Measurement of process-dependent subsurface damage in optical materials using the MRF wedge technique. *Proc. SPIE* **2006**, *5991*, 599103.

32. Cormont, P.; Gallais, L.; Lamaignère, L.; Donval, T.; Rullier, J.L. Effect of $CO_2$ laser annealing on residual stress and on laser damage resistance for fused silica optics. *Proc. SPIE* **2010**, *7842*, 78422C.

33. Jiang, X.; Liu, Y.; Rao, H.; Fu, S. Improve the laser damage resistance of fused silica by wet surface cleaning and optimized HF etch process. *Proc. SPIE* **2013**, *8786*, 87860Q.

34. Gao, X.; Feng, G.; Han, J.; Zhai, L. Investigation of laser-induced damage by various initiators on the subsurface of fused silica. *Opt. Express* **2012**, *20*, 22095–22101. [CrossRef] [PubMed]

35. Moncayo, M.A.; Santhanakrishnan, S.; Vora, H.D.; Dahotre, N.B. Computational modeling and experimental based parametric study of multi-track laser processing on alumina. *Opt. Laser Technol.* **2013**, *48*, 570–579. [CrossRef]

36. Wang, X.; Shen, Z.H.; Lu, J.; Ni, X.W. Laser-induced damage threshold of silicon in millisecond, nanosecond, and picosecond regimes. *J. Appl. Phys.* **2010**, *108*, 033103.

37. Samant, A.N.; Dahotre, N.B. Laser machining of structural ceramics—A review. *J. Eur. Ceram. Soc.* **2009**, *29*, 969–993. [CrossRef]

38. Timoshenko, S.P.; Goodier, J.N. *Theory of Elasticity*, 3rd ed.; McGraw-Hill Book Company: New York, NY, USA, 1970; pp. 1248–1297. 1297.

39. Lide, D.R. *Handbook of Chemistry and Physics*, 84th ed.; CRC Press: Boca Raton, FL, USA, 2003; pp. 27–80.

40. Martienssen, W.; Warlimont, H. *Handbook of Condensed Matter and Materials Data*, 1st ed.; Springer: New York, NY, USA, 2005; pp. 161–407, 523–559.

41. Walker, P.; Tarn, W.H. *Handbook of Metal Etchants*, 1st ed.; CRC Press: Boca Raton, FL, USA, 1991; pp. 76–106, 719–738.

42. Sato, K.; Yugami, H.; Hashida, T. Effect of rare-earth oxides on fracture properties of ceria ceramics. *J. Mater. Sci.* **2004**, *39*, 5765–5770. [CrossRef]

43. Miller, P.E.; Suratwala, T.I.; Wong, L.L.; Feit, M.D.; Menapace, J.A.; Davisand, P.J.; Steele, R.A. The Distribution of Subsurface Damage in Fused Silica. *Proc. SPIE* **2005**, *5991*, 599101.

44. Cheng, J.; Chen, M.; Liao, W.; Wang, H.; Xiao, Y.; Li, M. Fabrication of spherical mitigation pit on $KH_2PO_4$ crystal by micro-milling and modeling of its induced light intensification. *Opt. Express* **2013**, *21*, 16799–16813. [CrossRef] [PubMed]

45. Hu, G.; Zhao, Y.; Li, D.; Xiao, Q.; Shao, J.; Fan, Z. Studies of laser damage morphology reveal subsurface feature in fused silica. *Surf. Interface Anal.* **2010**, *42*, 1465–1468. [CrossRef]

*applied*
*sciences*

MDPI

*Review*

# Laser Welding under Vacuum: A Review

Meng Jiang, Wang Tao * and Yanbin Chen *

State Key Laboratory of Advanced Welding and Joining, Harbin Institute of Technology, Harbin 150001, China; 15B909076@hit.edu.cn
* Correspondence: taowang81@hit.edu.cn (W.T.); chenyb@hit.edu.cn (Y.C.); Tel.: +86-451-8641-5374 (W.T.); +86-451-8641-8645 (Y.C.)

Received: 10 August 2017; Accepted: 1 September 2017; Published: 5 September 2017

**Abstract:** Laser welding has been widely used in various industry fields. In order to further alter and broaden its applicability, a novel technology of laser welding under vacuum is introduced. The combination of high power laser and low ambient pressure provides an excellent welding performance and quality. In this paper, an overview on laser welding under vacuum is presented. It begins with a short introduction about the research status of laser welding under vacuum. Next, the equipment of laser welding under vacuum is introduced. Then, the fundamental phenomena of laser welding under vacuum, including penetration depth, weld geometry, plasma plume, molten pool and keyhole behaviors, are summarized in detail. Finally, the applications and prospects of laser welding under vacuum are proposed.

**Keywords:** laser welding under vacuum; ambient pressure; equipment; fundamental phenomena; application; prospect

## 1. Introduction

Welding, a joining method by heat or pressure to make the materials reach the connection between atoms, is the most versatile and realistic joining technology in every industrial field [1,2]. Like the arc, plasma and electron beams, a laser beam can also be used as heat source in the welding. Laser beam has been applied to welding since the first ruby laser was invented by Doctor Maiman [3–5]. In the last 20 years, laser welding has experienced rapid development. Because of the advantages of high quality, high precision, high efficiently, high performance, high flexibility, high speed, low distortion and low deformation, laser welding has been the most advanced and the best developing foreground welding method [6–9]. In the field of laser beam welding, constant attempts to alter and broaden the application possibilities have been made for years. New laser beam welding methods are appearing constantly, for example, laser welding with filler wire, remote laser welding, laser scanning welding, and so on [10–18]. Among these new laser welding methods, laser welding under vacuum or under reduced pressure is one of the most promising methods. Although there are some different opinions about laser welding under vacuum, the exceeding expected laser welding phenomena and excellent weld quality under vacuum or reduced pressure both attract the interests of researchers.

This paper makes a short overview of laser welding under vacuum. It begins with a short introduction about the research status of laser welding under vacuum. Next, the equipment of laser welding under vacuum are introduced. Then, the fundamental phenomena of laser welding under vacuum, including penetration depth, weld geometry, plasma plume, molten pool and keyhole behaviors, are presented in details. Finally, the applications and prospects of laser welding under vacuum are proposed.

## 2. The Research Status of Laser Welding under Vacuum

Laser welding under vacuum or reduced pressure has a history of more than 30 years. As shown in Figure 1, the first report about laser welding under vacuum or under reduced pressure dates back to the 1980s. Arata et al. [19] (Joining and Welding Research Institute, JWRI for short, at Osaka University) made the first effort to carry out the experiment of laser welding under vacuum conditions. The original intent of this research was to suppress the plasma and achieve a deep penetration in $CO_2$ laser welding. The problem of laser induced plasma, which is the most severe problem in laser welding at atmospheric, was completely solved by laser welding under vacuum. In addition, the exceeding expected penetration depth, which was approximate 2 times the depth of laser welding at atmospheric pressure, was achieved by laser welding under vacuum. However, this new laser welding method did not attract much attention due to the inadequate development of laser welding in industrial fields at that time. JWRI at Osaka University insisted on the research about laser welding under vacuum and paid their attention to fundamental phenomena in laser welding under vacuum. In 2001, the effect of vacuum on weld penetration and porosity formation was investigated in high-power $CO_2$ and YAG laser welding [20]. The reason for no porosity in vacuum was explained by keyhole and molten pool behaviors. In 2011, Katayama et al. [21] performed a high power disc laser welding experiment on 304 stainless steel and A5052 aluminum using a new chamber for laser welding under vacuum achieved by using rotary pumps. Sound deep single pass weld of 73 mm in penetration depth on type 304 stainless steel was obtained at a laser power of 26 kW, a welding speed of 0.3 m/min, a defocused distance of −40 mm and an ambient pressure of 0.1 kPa. The research of Katayama et al. [21] showed the penetrability of laser welding under vacuum which was similar to that of electron beam welding and the application feasibility of laser welding under vacuum on thick plate welding.

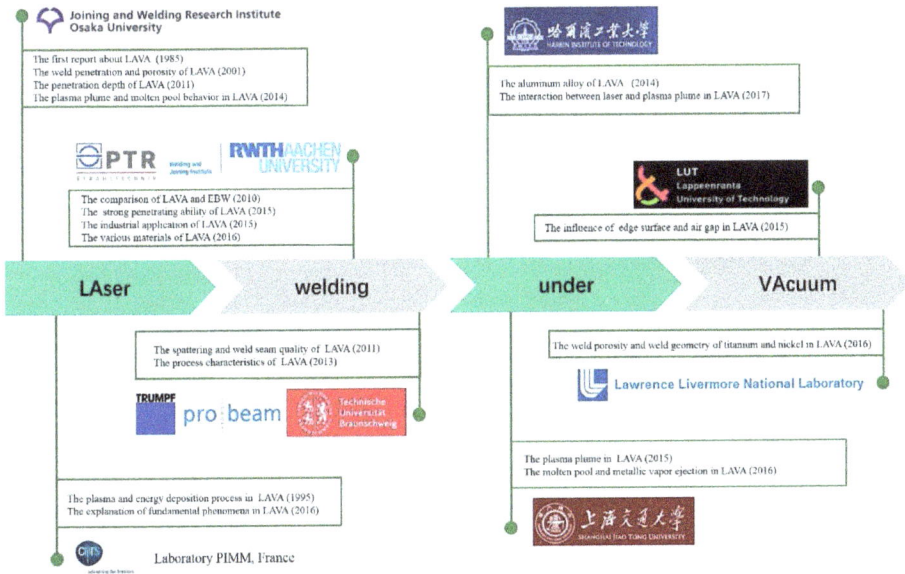

**Figure 1.** The research status of laser welding under vacuum.

Fabbro et al. (PIMM Laboratoire, Paris, France) have also been studying laser welding under vacuum or reduced pressure for many years. The research of Fabbro et al. mainly focused on the explanation of weld geometry changes and penetration improvement under vacuum conditions. In 1995, Fabbro et al. [22] observed the plasma suppression at low pressure by measuring plasma size, electron temperature and electron density. The energy deposition process inside the keyhole was compared to explain the different weld seam profiles between vacuum and atmosphere. Recently, Fabbro et al. [23] made a detailed explanation of fundamental phenomena in laser welding under vacuum by theoretical analysis and numerical simulation.

Since 2000, lasers have experienced an extremely rapid development. Especially during the last 10 years, the high-power (>10 kW, up to 100 kW) and high-brightness (<15 mm mrad) solid state lasers (fiber lasers or disk lasers) emerged into market [24,25]. Due to the availability of a new generation of high brightness multi-kilowatt solid state lasers, high-power laser beam welding became a new stimulus. However, some new problems, for example, the weld defects of sagging and spatter, the intense plasma plume and turbulent molten pool flow, are accompanied during high-power laser beam welding [26–28]. That means that the application of high power single pass laser welding encounters a bottleneck. Under such a background, the method of laser welding under vacuum or reduced pressure has been enjoying a renaissance.

In 2010, the Welding and Joining Institute (ISF) of RWTH Aachen University started research work of laser welding under vacuum using a 600 W single-mode fiber laser and a 12 kW multimode fiber laser [29]. The weld formations of both thick plate and thin plate were compared between laser welding under vacuum and electron beam welding. Based on the impressive results of the preliminary study, ISF of RWTH Aachen University built up a welding equipment for laser welding under vacuum on cooperation with PTR Strahltechnik GmbH A large number of process exploration tests of laser welding under vacuum were carried out on this new laser welding equipment [30–32]. The results of process tests showed that laser welding had a strong penetrating ability and a good adaptability for materials. Various materials, for example unalloyed steel, high alloyed steel, nickel-base alloy, titanium alloy and even copper alloy, were welded and achieved a deep and sound weld. In terms of the research results of laser welding under vacuum, ISF of RWTH Aachen University actively promoted this technology in the industrial field.

In 2011, Institute of Joining and Welding (ifs) at TU Braunschweig started studying laser welding under vacuum on a corporation with Trumpf Laser- und Systemtechnik GmbH and Pro-beam M [33,34]. Instead of the beam generator, a laser beam was coupled into a vacuum chamber of a former electron beam welding machine to carry out laser welding under vacuum. The influence of ambient pressure on spattering and weld quality in laser beam welding was presented. The process characteristics of laser welding at reduced ambient pressure was studied in detail.

In recent years, more and more researchers have shown their interest in laser welding under vacuum. In 2014, the State Key Laboratory of Advanced Welding and Joining (AWJ) at Harbin institute of technology investigated the laser bead on plate welding of aluminum alloy under reduced pressure [35]. In 2017, the effect of ambient pressure on weld formations and the interaction between laser radiation and plasma plume were further investigated [36]. In 2015, Laboratory of Laser Processing at Lappeenranta University of Technology studied the influence of joint edge surface edge roughness and pre-set air gap on the weld quality and penetration depth in partial vacuum conditions laser beam welding [37]. In 2014, Shanghai Jiao Tong University investigated the plasma plume in fiber laser welding under subatmospheric pressure [38]. In 2016, the dynamic coupling between molten pool and metallic vapor ejection in fiber laser welding under subatmospheric pressure was further studied [39]. In 2015, Daimler AG in Germany carried out the welding experiments of 16MnCr5 steel and ALMg3 aluminum using X-ray analysis to investigate the keyhole behavior at the reduced ambient pressure [40]. In 2016, Lawrence Livermore National Laboratory, USA, performed the study about the effect of reduced pressure on laser keyhole weld porosity and weld geometry in pure titanium and nickel [41].

Significant advantages of laser welding under vacuum or reduced pressure are attracting the attention of researchers and show great value of research and practical application. In the further 5–10 years, laser welding under vacuum will become a key issue in laser beam welding.

## 3. The Equipment of Laser Welding under Vacuum

In order to carry out laser welding under vacuum, the first issue is to build up a vacuum chamber for laser welding. The simplest way is to refer to the basic design of electron beam welding. Instead of an electron beam generator, a laser processing head is installed in vacuum chamber. The design of a built-in laser head has the advantage of a more flexible laser welding under vacuum machine. However, a special seal should be designed to couple optical fiber and water-cooled pipelines into vacuum chamber. With the cooperation of ISF at RWTH Aachen University and II-VI HIGHYAG, a fiber-fiber-coupler was design to develop an easy and affordable method to transfer laser power into the vacuum. The fiber-fiber-coupler open the possibility of high power laser processing with built-in laser head under vacuum [42]. Unlike electron beams, laser radiation is a matter of light. Laser beams can pass through the glass into a vacuum. Therefore, there is another structural design for laser welding under vacuum. Laser processing head is fixed outside the vacuum chamber. Only the laser beam is coupled into the vacuum chamber through glass. As shown in Figure 2, most of the research institutions used the second design according to the current reports. JWRI at Osaka University created a transparent vacuum chamber using acrylic [43,44]. The rotary pump was used to achieve vacuum conditions. The design of transparent vacuum chamber was beneficial to take photos with high speed camera and other optical tests. AWJ of Harbin Institute of Technology set up a $680 \times 400 \times 400$ mm vacuum chamber. The theoretical limit vacuum level was $6.6 \times 10^{-4}$ Pa by using a rotary pump and a molecular pump simultaneously. The ifs at TU Braunschweig reformed a 1.5 m$^3$ vacuum chamber of former electron beam welding plant [33]. A blank flange was manufactured in which a quartz glass was located as a coupling-in window to couple laser into chamber. ISF at RWTH Aachen University established a cylinder vacuum chamber of 0.6 m$^3$ [31]. The vacuum chamber was equipped with two rotary vacuum pumps. This vacuum chamber had a minimum vacuum pressure of $10^{-3}$ hPa and the ability to reach 0.1 hPa in 5 min. Besides, the vacuum chamber could realize different welding positions by turning around the longitudinal axis.

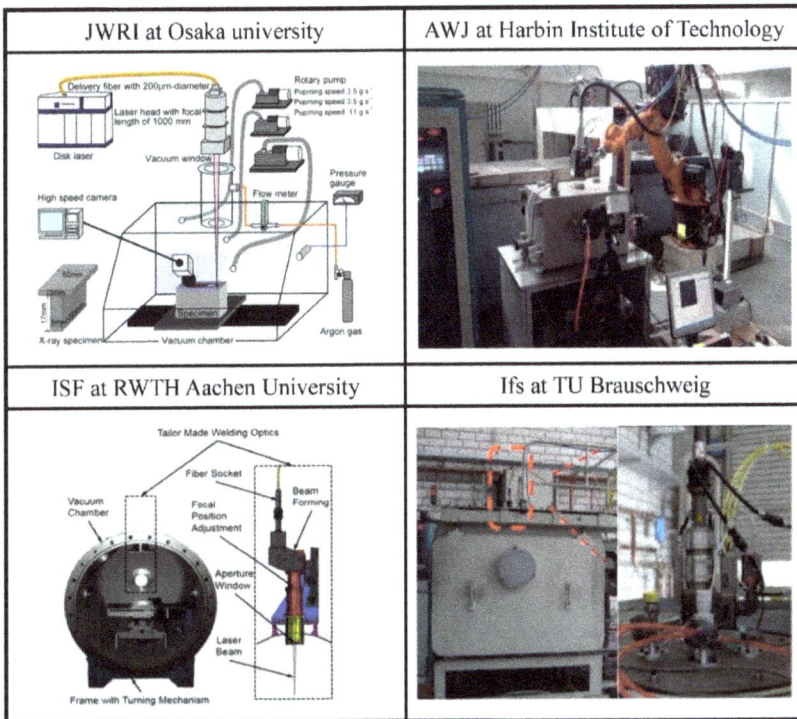

**Figure 2.** The equipment of laser welding under vacuum. Reproduced with permission from [31], Springer, 2016; Reproduced with permission from [33], Laser Institute of America, 2011; Reproduced with permission from [44], Taylor & Francis, 2014.

According to the above designs of vacuum chamber, the equipment of laser welding under vacuum basically consists of three parts: laser processing head, vacuum chamber and coupling-in window. The laser processing head should be set above the vacuum chamber. A conventional system for laser welding at atmosphere can easily fulfill the requirement. Compared with electron beam welding, laser welding under vacuum has a relatively low requirement of vacuum level. Therefore, the vacuum chamber structure of laser welding is relatively simple. Hence, the key part of the vacuum chamber is the coupling-in window for coupling the laser beam into the vacuum. A ZnSn glass were used as coupling-in window in $CO_2$ laser welding under vacuum. A quartz glass was used for solid laser welding under vacuum. In laser welding under vacuum, intense metal vapor ejected from keyhole and deposited on the surface of glass. That means that the stability of laser propagation and the lifetime of glass could be seriously affected. A laser head with long focusing length [31,43,44] (JWRI at Osaka University and ISF at RWTH Aachen University) or a special coupling-in window [32,34] (ISF at RWTH Aachen University and ifs at TU Braunschweig) with a rather small gas flow under the glass are used to protect the coupling-in window from process emissions. Furthermore, the coupling-in glass is coated with an antireflection film and equipped with a water-cooling device for long-time welding with high power.

In the last 2–3 years, the systematic equipment of laser welding under vacuum entered the market. As shown in Figure 3, LAVA-L95 and LASVAC PL 01 are two typical systematic equipment of laser welding under vacuum [45–47]. With the support of ISF at RWTH Aachen University, the machines were developed separately by FOCUS Gmbh and PTR Company. In order to meet the demand for different processes or products, a unit assembly system is designed. That means the different laser sources and optics can be chosen in the equipment of laser welding under vacuum.

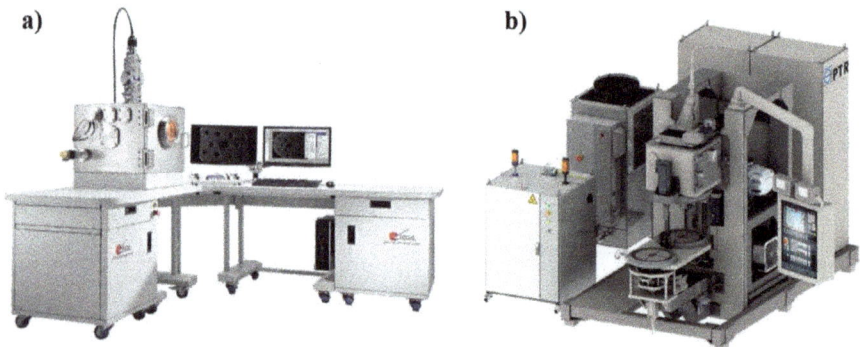

**Figure 3.** The systematic equipment of laser welding under vacuum (**a**) LAVA-L95. Reproduced with permission from [45], FOCUS electronics GmbH, 2017. (**b**) LASVAC PL 01. Reproduced with permission from [47], PTR Strahltechnik GmbH, 2017.

## 4. Fundamental Phenomena during Laser Welding under Vacuum

### 4.1. The Influence of Ambient Pressure on Penetration Depth and Weld Geometry

The increase in penetration depth and change in weld geometry are the most obvious achievements brought by the reduction of ambient pressure. According to the present experiment results [31,33,36,38,44], the penetration depth increases with the decrease of the ambient pressure. Based on the effect of ambient pressure on penetration depth, there exists a critical threshold of ambient pressure. That means the penetration no longer increases, and sometimes probably decreases when the ambient pressure is lower than the critical threshold. Because of the different machines and different experimental conditions, the critical pressures are not consistent. The previous results [31,33,36,38,44] showed that the critical pressure was in the range of 0.1 kPa to 10 kPa. Abe et al. [44] carried out the bead-on-plate welding of type 304 stainless steel plates and A5052 aluminum alloy with a 16 kW laser power at a welding speed of 1 m/min under reduced pressure. The results showed that the critical pressure was about 10 kPa and the penetration depth of both type 304 stainless steel plates and A5052 aluminum alloy at 10 kPa is nearly twice as deep as that at 101 kPa. The results of Abe et al. [44] also revealed that the weld width decreased with the decreasing ambient pressure. The increase in penetration depth and the decrease in weld width led to a large difference in aspect ratio (penetration depth/weld width). That meant that the weld geometry between under vacuum and atmosphere were totally different. In comparison with conventional laser welding at atmospheric pressure, a deep and parallel weld, which is similar to electron beam welding, is achieved during laser welding under vacuum.

The results of ISF at RWTH Aachen university also proved the strong penetrating ability of laser welding under vacuum [31,48]. The cross section profiles of laser welding at atmosphere, laser welding under vacuum and electron beam welding are shown in Figure 4. With the same heat input, the penetration depth of both laser welding under vacuum and electron beam welding is about 50 mm. However, the pressure used in electron beam welding is two orders of magnitude lower than in laser welding under vacuum. The strong penetrating ability under low vacuum is the unique advantage of laser welding under vacuum.

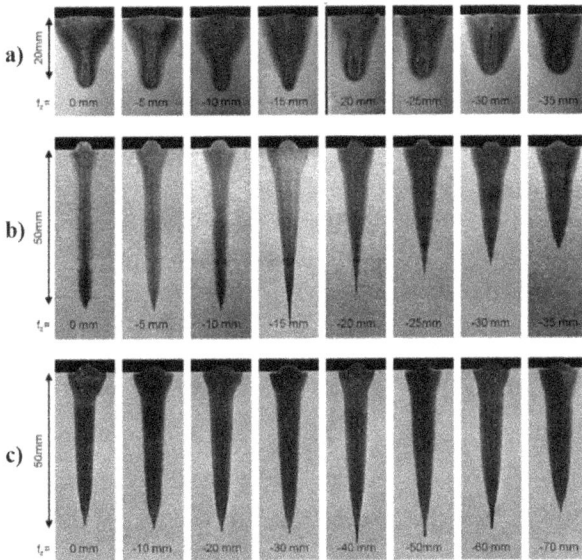

Figure 4. Comparison of cross section profiles using different welding processes (a) Laser welding (laser power is 16 kW, welding speed is 0.3 m/min, ambient pressure is 1000 mbar), (b) Laser welding under vacuum (laser power is 16 kW, welding speed is 0.3 m/min, ambient pressure is $10^{-1}$ mbar), (c) Electron beam welding (electron beam power is 16 kW, welding speed is 0.3 m/min, ambient pressure is $10^{-3}$ mbar). Reproduced with permission from [48], John Willey and Sons, 2015.

The ambient pressure, welding speed and defocused distance are the main parameters that control the weld appearance and penetration depth during laser welding under vacuum. It was found that, under vacuum condition, the increase of penetration depth was quite different at various welding speed [33,49]. As shown in Figure 5 (Figure 6 of "Influence of ambient pressure on spattering and weld seam quality in laser beam welding with the solid-state laser" reproduced with permission from Proceedings of the International Congress on Applications of Lasers and Electro-optics, October 2011), a remarkable increase of penetration depth occurs at low to medium speeds (up to approx. 3.0 m/min), it is true that there is a limit in the ambient pressure for improving penetration depth. Besides the lower the welding speed is, the more obvious the effect of ambient pressure on penetration depth is. When the welding speed is over 4 m/min, the ambient pressure has almost no impact on the penetration depth. The laser focus position carries with the most concentrated laser energy. The focus position (above the workpiece or below the workpiece) has an influence on the laser energy transfer along the keyhole. In the case of laser welding under vacuum, the laser energy disposition along the extreme deep keyhole has a distinct difference compared with the conventional laser welding. The keyhole depth under vacuum is much deeper than that at atmosphere. Therefore, the defocused distance is an important parameter to control the weld appearance and penetration depth under vacuum. In addition, the oscillating laser beam has been introduced into laser welding under vacuum. This has proved that the oscillation of laser beam affected the weld seam geometry, spattering behavior and weld meatal ajections [50].

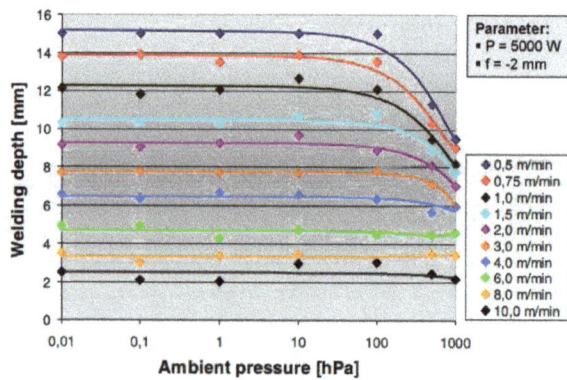

**Figure 5.** Effect of ambient pressure on penetration depth at various welding speed ICALEO® 2011 Proceedings. Reproduced with permission from [33], Laser Institute of America, 2011.

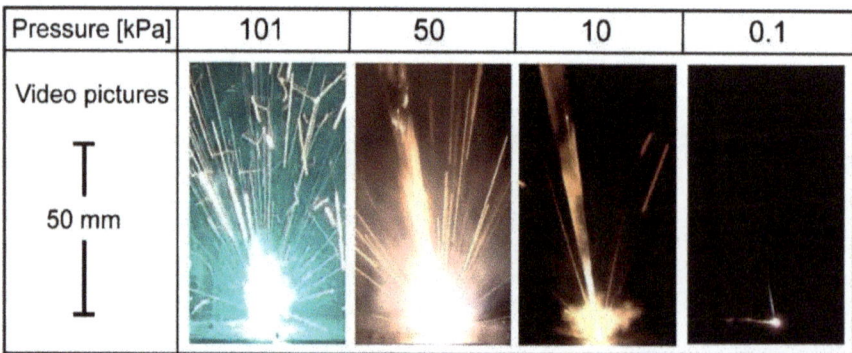

**Figure 6.** Typical video pictures of A5052 aluminum alloy laser induced plume observed under each reduced pressure (laser power is 16 kW, welding speed is 1 m/min, defocus distance is 0 mm). Reproduced with permission from [44], Taylor & Francis, 2014.

### 4.2. The Influence of Ambient Pressure on Plasma Plume

It was known that the welding vapor plasma exerted a negative "shielding effect" in high-power $CO_2$ laser welding. The strong inverse bremsstrahlung (I.B.) effect decreases the penetration depth and even stops the $CO_2$ laser welding process due to the optical breakdown [51,52]. The vapor plasma exerts a different state between $CO_2$ laser welding and solid laser welding. Due to the presence of a large number of non-ionized metal particles, the vapor plasma is called "metal vapor plume" in solid laser welding [53]. However, the plasma plume still causes some disturbances by Rayleigh or Mie scattering in solid laser welding [54–56]. The previous studies [19,33,44] showed that no matter whether $CO_2$ laser welding or solid laser welding was used, the reduction of ambient pressure had a significant effect on plasma plume behavior. As shown in Figure 6 an extremely intensive and very bright plasma plume was exerted above the workpiece in high power laser welding at atmospheric pressure (101 kPa). The lower part plasma plume adjoined directly the keyhole has much higher brightness than the other part. Besides, a large number of spatters were observed around the plasma plume. The reduction of ambient caused an obvious change in plasma plume. When the ambient pressure reduced to 50 kPa, the change of plasma plume was already visible. Both the intensive luminescence and the volume of plasma plume decreased. At the reduced ambient pressure of 10 kPa, the intense illuminant plasma plume and obvious spattering disappeared. There was only a small brightness and volume of plasma plume above the keyhole. When the ambient pressure decreased at 0.1 kPa, the plasma plume was completely suppressed and invisible. The different materials and the welding parameters led the difference in critical ambient pressure of plasma plume disappearance.

The enormous difference in the luminescence and the volume of plasma plume guide the research of the interaction between laser radiation and plasma plume under different ambient pressures. As shown in Figure 7, a probe laser, which horizontally passed through the plasma plume, was usually used to study the interaction between laser radiation and plasma plume [54,56]. As shown in Figure 8, the spot behaviors of a probe laser beam passing through the plasma plume under various ambient pressures are presented [36]. The results showed that the interaction between laser radiation and plasma plume was minimal at low ambient pressures. Therefore, the laser process was more stable under vacuum. In addition, the extinction of fiber laser radiation decreased with decreasing ambient pressure [38]. Although the interaction between laser radiation and plasma plume under different ambient pressures has been discussed, the essential influence of ambient pressure on plasma plume has not been presented in detail. On the one hand, the illuminant change of plasma plume means that the ambient pressure has a significant effect on the ionization state and formation condition of the plasma plume. On the other hand, the ambient pressure has a direct relation with the stress state of metal vapor behavior. This may lead to a different laser welding process.

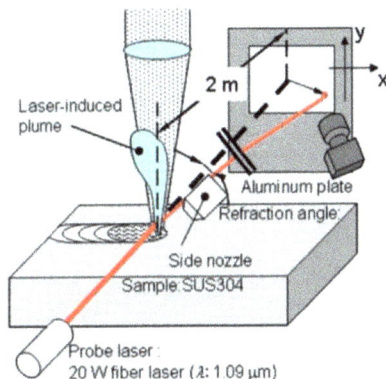

**Figure 7.** The schematic of studying the interaction between laser radiation and plasma plume using a probe laser. Reproduced with permission from [54], Taylor & Francis, 2008.

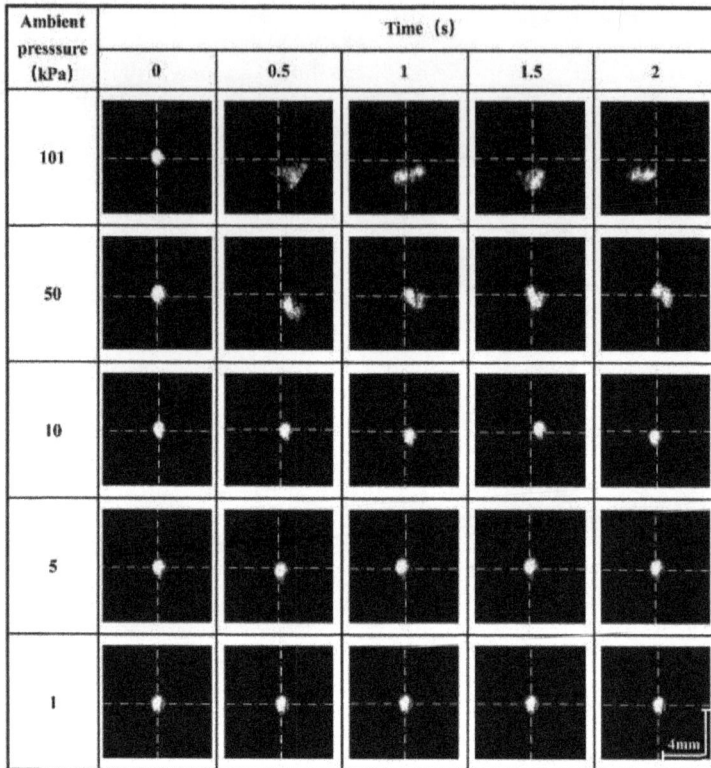

**Figure 8.** The spot behaviors of probe laser beam passing through the plasma plume under various ambient pressures (laser power is 5 kW, welding speed is 0.25 m/min, defocus distance is 0 mm). Reproduced with permission from [36], Elsevier, 2017.

*4.3. The Influence of Ambient Pressure on Molten Pool and Keyhole Behaviors*

During laser welding, the molten pool and keyhole behaviors have a close relationship with weld quality and defect formation. Due to the specialty of molten pool and keyhole behaviors, it is difficult to understand and investigate the molten pool and keyhole behaviors. Therefore, the advanced imaging technology and numerical simulation are the main methods to study the molten pool and keyhole behaviors. Both methods are used to explore the effect of ambient pressure on the molten pool and keyhole behaviors. The surface molten pool and keyhole inlet of SUS 304 stainless steel under different ambient pressures were observed by Youhei et al. [44]. As shown in Figure 9, the surface molten pool and the keyhole inlet changed following the ambient pressure decreasing. The average diameter of keyholes decreased, the surface molten pool became narrower and more stable under vacuum conditions. Moreover, there were quite different liquid flows between atmospheric pressure and vacuum. Katayama et al. [20] presented the liquid flow in the molten pool at various ambient pressures using the micro focused X-ray real-time observation. Based on the results of Katayama et al. [20], there were two main differences in liquid flow between atmosphere and vacuum. On the one hand, the liquid flowed downward along the rear keyhole at atmospheric pressure. While the liquid flowed upwards along the rear keyhole wall under vacuum. That means the liquid flow along the rear keyhole is opposite. On the other hand, there was a molten flow along the bottom molten pool at atmospheric pressure, while there was no such strong flow near the bottom under vacuum. The differences in molten pool flow had a direct relationship with the porosity formation. In addition, the keyhole shapes at different pressures were observed directly by Engelhardt et al. [40]. Figure 10 (Figure 5 of "Time-resolved X-ray Analysis of the Keyhole Behavior during Laser Welding of Steel and Aluminum at Reduced Ambient Pressure" reproduced with permission from Proceedings of the International Congress on Applications of Lasers and Electro-optics, October 2015) shows the averaged keyhole shapes at ambient pressures from 0.5 kPa to 101 kPa. The keyhole depth increased with the decreasing of ambient pressure. This phenomenon corresponded to the increase in penetration depth under vacuum. In addition, the reduction of ambient pressure resulted in an increased inclination angle of the keyhole front and an increased bending of the keyhole tip.

| Pressure [kPa] | 101 | 50 | 10 | 0.1 |
|---|---|---|---|---|
| High-speed picture 1 mm ⌐1 mm | | | | |
| Schematic illustration | | | | |
| Average size of keyhole inlet [mm] | 1.2 | 1.1 | 0.7 | 0.7 |
| Average width of molten pool [mm] | 1.8 | 3.1 | 1.5 | 0.9 |

**Figure 9.** High speed video picture of keyhole inlet and molten pool in type 304 stainless steel weld observed under each reduced pressure(laser power is 16 kW, welding speed is 1 m/min, defocus distance is 0 mm). Reproduced with permission from [44], Taylor & Francis, 2014.

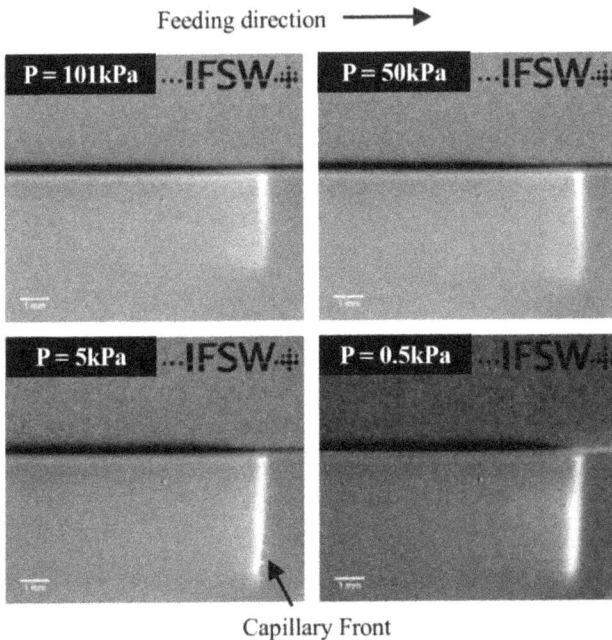

**Figure 10.** X-ray images of averaged keyhole shapes at ambient pressures from 0.5 kPa–101 kPa (laser power is 2 kW, welding speed is 1.2 m/min, defocus distance is 0 mm). ICALEO® 2015 Proceedings. Reproduced with permission from [40], Laser Institute of America, 2015.

Pang et al. [23,57,58] investigated the physical phenomenon of laser welding under vacuum by numerical simulation. Based on the 3D transient multiphase model of laser welding, an improved model of recoil pressure for laser welding under any ambient pressure was proposed. The comparison of the predicted keyhole wall temperature and the predicted velocity distribution of metallic vapor under atmosphere and vacuum are presented. The results show that the average keyhole wall temperature became lower and the predicted speed of metallic vapor increased under vacuum. Based on the previous results of numerical simulation, Fabbro et al. [23] presented that the penetration depth improvement in laser welding under vacuum mainly resulted from the reduction of evaporation temperature. Less power per unit depth of keyhole is necessary due to the reduction of evaporation temperature, resulting in a deeper keyhole with the same incident laser power. The saturation of the penetration depth below some critical pressure is related to the recoil pressure. In addition, the recoil pressure and welding speed are positively related. At high welding speed, the recoil pressure is higher. Moreover, it is much higher than ambient pressure. Therefore, the reduction of ambient pressure cannot change the evaporation pressure inside the keyhole. This is the reason that at high welding speeds there is no improvement in penetration depths while at low welding speed, the reduction of ambient pressure will modify the evaporation pressure inside the keyhole due to relatively low recoil pressure. The low evaporation pressure results in the improvement in penetration depths under vacuum.

## 5. Applications of Laser Welding under Vacuum

As an emerging welding technology, laser welding under vacuum has received much attention in recent years. The reports of laser welding under vacuum mainly focused on the effect of ambient pressure on laser welding phenomena. Laser welding under vacuum is still at the beginning stages of application. Based on the advantages of laser welding under vacuum, the applications of this competitive technology are expected. The main application attempt was on the mass production of drive section components with radial and/or axial weld seams [47,59,60]. Within the framework of a cooperation of PTR Company and RWTH Aachen University, a cycle machine for welding with laser beam under vacuum was implemented. The planet wheel carrier and gear wheel were fabricated by the cycle machine for laser welding under vacuum. As shown in Figure 11, the plate wheel carriers consist of a planet wheel carrier with axle arm and a coronoid bole carrier. The welding zone is triangular and the maximally required penetration depth is 25 mm. A sound welded joint without oxides and spatters was fabricated using a laser power of 7.5 kW, a welding speed of 0.42 m/min, a focus position of −5 mm and an ambient pressure of 2 kPa. At the same time, Pro-beam, TRUMPF and the ifs at TU Braunschweig are also adapting the process for the low-pressure, spatter-free laser welding of powertrain components into industrial application [61,62]. The powertrain components requiring weld penetration depths of between three and six millimeters and a high feed rate were manufactured. Because of stable process without spatter, high weld quality and proper penetration depth, laser welding under vacuum is very suitable for the fabrication of powertrain components.

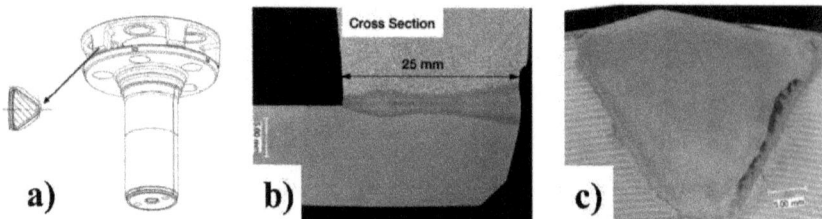

**Figure 11.** The planet wheel carrier welded by laser welding under vacuum ((laser power is 7.5 kW, welding speed is 0.42 m/min, defocus distance is −5 mm, ambient pressure is 20 mbar-1). Reproduced with permission from [59], John Willey and Sons, 2015. (**a**) The schematic of planet wheel carrier, (**b**) Cross-section in stud center, (**c**) Longitudinal section through the weld.

Laser welding has been widely used in various productions. However, laser welding is mainly used in the sheet structure with small thickness. With the advent of new high power solid lasers, the use of solid-state laser in deep penetration welding has been constantly explored. The exceeding expected phenomena of laser welding under vacuum make it very suitable for thick plate applications where low welding speeds can be applied to achieve very high penetration depth and for welding tasks with the highest demands on the weld quality. Some various materials, such us copper alloys, structural steels, duplex stainless steels, nickel-based alloys, titanium alloys and so on, have been welded by ISF at RWTH Aachen University [30–32,46]. As shown in Figure 12, the plate thickness of 50 mm for unalloyed steel S690QL was achieved by single pass laser welding under vacuum at a laser power of 16 kW, a welding speed of 0.5 m/min, a focus position of −5 mm and an ambient pressure of 10 Pa. The full penetration weld joint of 38 mm nickel-base alloy without was fabricated at a laser power of 16 kW, a welding speed of 0.4 m/min, a focus position of 0 mm and an ambient pressure of 10 Pa. As shown in Figure 12, both the weld joints presented a sound weld with super deep penetration depth. It is clear that laser welding under vacuum has the developing prospect of welding heavy section components under low vacuum. As we know, electron beam welding is thought to be suitable for single pass weld of thick plate [29,63]. In this regard, laser welding under vacuum is able to challenge the electron beam welding in terms of the weld quality and penetration depth. In addition, laser welding under vacuum has the following advantages: relatively low vacuum level, lack of X-ray radiation protection and possible arising of nonmetallic [29,64].

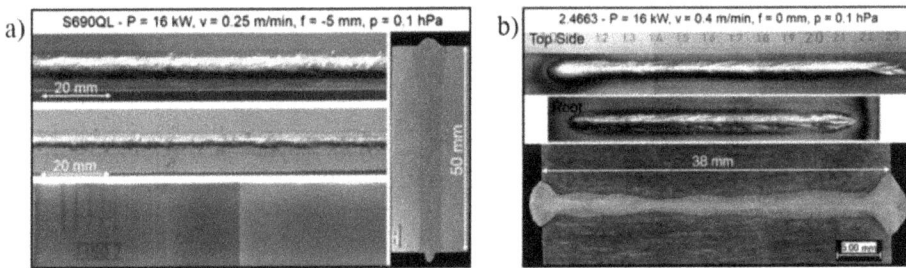

**Figure 12.** The survey of weldable plate thicknesses/materials with laser welding under vacuum. Reproduced with permission from [31], Springer, 2016. (**a**) Laser welding under vacuum of unalloyed steel S690QL, (**b**) Laser welding under vacuum of nickel-base alloy 617.

## 6. Prospects of Laser Welding under Vacuum

With the reference to the previous researches of laser welding under vacuum, the main advantages of laser welding under vacuum or reduced pressure are as follows [45,47,61,62]:

- Significant increase (More than two times) of the welding depth;
- Parallel-sided seams with reduced nail head;
- Increased process stability due to greatly reduced and stable plasma plume;
- Reduction of workpiece contamination by spatter and vaporization;
- Higher-quality, pore-free weld seams;
- Low operating costs due to efficient solid-state laser;
- Welding process without inert gas;
- Lower vacuum level and simpler device compared with electron beam welding.

Despite the advantages of laser welding under vacuum above, the vacuum chamber causes laser welding loose the most important advantage: flexibility. In addition, the dimensions of the welded part were restricted by the dimensions of the vacuum chamber. A larger vacuum chamber means higher cost and associated pumping time [65]. A lot of trails were carried out to make the electron beam welding release the limitation of vacuum chamber. Rolls-Royce and TWI worked on a local vacuum, or "out of chamber" to expand the application of electron beam welding in thick section components [66]. However, the electron beam interaction with low pressure gases is the difficult point in the application of electron beam welding under local vacuum. However, the interaction of the laser with ambient gases is negligible. In addition, the relatively low vacuum can achieve an obvious increase in penetration depth. Therefore, laser welding under local vacuum or mobile vacuum has a more practical application value. ISF RWTH-Aachen and BAM, Germany has conducted some attempts in laser welding under mobile vacuum [66–68]. As shown in Figure 13, a vacuum cap, which provides a local reduced ambient pressure of about 20 kPa above the welding area, was presented by BAM. The reduced pressure in the vacuum cap generate 50% higher penetration depth in comparison to welding under ambient pressure conditions. Due to the restriction of this movable seal, the vacuum level is very low and the penetration depth has no significant increase. The laser welding under vacuum can be combined with local vacuum electron beam welding to accumulate the disadvantages of those two processes. Hence, a local vacuum seal with a laser, instead of the electron beam generator may be a promising technique in the future.

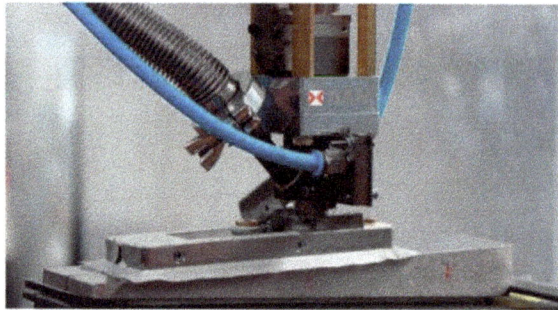

**Figure 13.** The mobile local vacuum chamber for high power laser beam welding of thick materials. Reproduced with permission from [66], John Willey and Sons, 2015.

Laser welding under vacuum not only has promising application prospects, but also provides a special physical environment to achieve a fundamental understanding of the laser welding process. Laser keyhole welding is a complex physical process involving heat flow, melting, evaporation, solidification phase changes, melt flow, and vapor flow. Most physical phenomena in laser keyhole welding have not been fully understood due to the complexity of laser welding. The phenomena of laser welding are completely different at the same welding parameter except for ambient pressure. By comparing the differences between atmosphere and vacuum and analyzing the influences of ambient pressure, the essences of laser welding can be more deeply understood. For example, the interaction of laser radiation and plasma plume is a focused issue in high power laser welding. Under vacuum conditions, the plasma plume was suppressed completely. By studying the effect of ambient pressure on the plasma plume, we could get a better understanding of the interaction of laser radiation and plasma plume.

There is no such thing as a perfect technology. The laser welding under vacuum is no exception. There are also some challenges in laser welding under vacuum. As is stated above, the coupling-in window for coupling laser beam into vacuum is the key part of vacuum chamber for laser welding. The lifetime of coupling-in window determines the stability of the equipment of laser welding under vacuum. Besides, higher and higher laser power will be used in laser welding under vacuum in the future. The coupling-in window glasses should have a higher maximum limiting power. As we know, the laser processing head itself consists of optical glass. In the future, the integration of a laser processing head and coupling-in window on the vacuum chamber is probably a development trend of equipment of laser welding under vacuum. On the other hand, there are also many mechanisms yet to be explained in laser welding under vacuum. The research about the effect of ambient pressure on the plasma ionization and vapor ejection in plasma plume, the dynamic behavior of keyhole and molten pool, and the metallurgical behavior under vacuum will provide a good deal of insight into the fundamental processes of laser welding under vacuum.

## 7. Conclusions

A review of laser welding under vacuum including histories, equipment, fundamental phenomena, applications and prospects has been reported. The first test of laser welding under vacuum or under reduced pressure dates back to the 1980s. The development of high-power lasers means laser welding under vacuum is enjoying a renaissance. Compared with electron beam welding, the equipment of laser welding under vacuum is relatively simple. The coupling-in window for coupling the laser beam into a vacuum is the key part of vacuum system for laser welding. The ambient pressure has a significant influence on the phenomena of laser welding. The changes of plasma plume, keyhole and molten pool behaviors result in an excellent sound and deep welding performance. Based on the advantages of laser welding under vacuum, the application on drive section and powertrain components has been attempted in Germany. A local mobile vacuum for laser welding is expected to be a promising technology in the future. Besides, the vacuum environment is a benefit to further understanding the physical process of laser keyhole welding.

**Acknowledgments:** This work was supported by the National Key Research and Development Program of China (2016YFB1102100), "the Fundamental Research Funds for the Central Universities" (Grant NO.HIT.NSRIF.2017003) and the Nature Science Foundation of Heilongjiang Province (E2016027).

**Conflicts of Interest:** The authors declare no conflict of interest.

## References

1.  Kou, S. *Welding Metallurgy*; Wiley: New York, NY, USA, 1987.
2.  Lancaster, J.F. *Metallurgy of Welding*; Elsevier: Amsterdam, The Netherlands, 1999.
3.  Katayama, S. *Handbook of Laser Welding Technologies*; Elsevier: Amsterdam, The Netherlands, 2013.
4.  Allmen, M.V.; Blatter, A. *Laser-Beam Interactions with Materials: Physical Principles and Applications*; Springer Science & Business Media: Berlin, Germany, 2013.
5.  Duley, W.W. *Laser Welding*; Wiley: Hoboken, NJ, USA, 1999.
6.  Kacar, I.; Ozturk, F.; Yilbas, B.S. A review of and current state-of-the-art in laser beam welding in the automotive industry. *Laser Eng.* **2016**, *33*, 327–338.
7.  Cao, X.; Jahazi, M.; Immarigeon, J.P.; Wallace, W. A review of laser welding techniques for magnesium alloys. *J. Mater. Process. Technol.* **2006**, *171*, 188–204. [CrossRef]
8.  Martukanitz, R.P. A critical review of laser beam welding. In *Lasers and Applications in Science and Engineering, Proceedings of the Society of Photo-Optical Instrumentation Engineers (SPIE), San Jose, CA, USA, 22–27 January 2005*; Schriempf, J.T., Ed.; SPIE: Bellingham, WA, USA, 2005; Volume 5706, pp. 11–24.
9.  Cao, X.; Wallace, W.; Poon, C.; Immarigeon, J.P. Research and Progress in Laser Welding of Wrought Aluminum Alloys. I. Laser Welding Processes. *Mater. Manuf. Process.* **2003**, *18*, 1–22. [CrossRef]
10. Dilthey, U.; Fuest, D.; Scheller, W. Laser welding with filler wire. *Opt. Quantum Electron.* **1995**, *27*, 1181–1191.
11. Salminen, A.S.; Kujanpää, V.P. Effect of wire feed position on laser welding with filler wire. *J. Laser Appl.* **2003**, *15*, 2–10. [CrossRef]
12. Salminen, A.S. Effects of filler wire feed on the efficiency of laser welding. In *LAMP 2002: International Congress on Laser Advanced Materials Processing, Proceedings of SPIE—The International Society for Optical Engineering, Osaka, Japan, 27–31 May 2002*; SPIE: Bellingham, WA, USA, 2003; pp. 263–268.
13. Rasmussen, D.; Dubourg, L. Hybrid laser-GMAW welding of aluminum alloys: A review. In Proceedings of the 7th International Conference on Trends in Welding Research, Pine Mountain, Atlanta, GA, USA, 16–20 May 2005; pp. 133–142.
14. Bagger, C.; Olsen, F.O. Review of laser hybrid welding. *J. Laser Appl.* **2005**, *17*, 2–14. [CrossRef]

15. Lu, J.; Kujanpää, V. Review study on remote laser welding with fiber lasers. *J. Laser Appl.* **2013**, *25*, 052008. [CrossRef]
16. Reinhart, G.; Munzert, U.; Vogl, W. A programming system for robot-based remote-laser-welding with conventional optics. *CIRP Ann. Manuf. Technol.* **2008**, *57*, 37–40. [CrossRef]
17. Hatwig, J.; Minnerup, P.; Zaeh, M.F.; Reinhart, G. An automated path planning system for a robot with a laser scanner for remote laser cutting and welding. In Proceedings of the 2012 IEEE International Conference on Mechatronics and Automation (ICMA), Chengdu, China, 5–8 August 2012; pp. 1323–1328.
18. Hao, K.; Li, G.; Gao, M.; Zeng, X. Weld formation mechanism of fiber laser oscillating welding of austenitic stainless steel. *J. Mater. Process. Technol.* **2015**, *225*, 77–83. [CrossRef]
19. Arata, Y.; Abe, N.; Oda, T. Fundamental Phenomena in High Power CO_2 Laser (Report II): Vacuum Laser Welding (Welding Physics, Process & Instrument). *Trans. JWRI* **1985**, *14*, 217–222.
20. Katayama, S.; Kobayashi, Y.; Mizutani, M.; Matsunawa, A. Effect of vacuum on penetration and defects in laser welding. *J. Laser Appl.* **2001**, *13*, 187–192. [CrossRef]
21. Katayama, S.; Yohei, A.; Mizutani, M.; Kawahito, Y. Development of Deep Penetration Welding Technology with High Brightness Laser under Vacuum. *Phys. Procedia* **2011**, *12*, 75–80. [CrossRef]
22. Verwaerde, A.; Fabbro, R.; Deshors, G. Experimental study of continuous $CO_2$ laser welding at subatmospheric pressures. *J. Appl. Phys.* **1995**, *78*, 2981–2984. [CrossRef]
23. Fabbro, R.; Hirano, K.; Pang, S. Analysis of the physical processes occurring during deep penetration laser welding under reduced pressure. *J. Laser Appl.* **2016**, *28*, 022427. [CrossRef]
24. Bachmann, M.; Gumenyuk, A.; Rethmeier, M. Welding with High-power Lasers: Trends and Developments. *Phys. Procedia* **2016**, *83*, 15–25. [CrossRef]
25. Nielsen, S.E. High Power Laser Hybrid Welding—Challenges and Perspectives. *Phys. Procedia* **2015**, *78*, 24–34. [CrossRef]
26. Avilov, V.V.; Gumenyuk, A.; Lammers, M.; Rethmeier, M. PA position full penetration high power laser beam welding of up to 30 mm thick AlMg3 plates using electromagnetic weld pool support. *Sci. Technol. Weld. Join.* **2012**, *17*, 128–133. [CrossRef]
27. Kawahito, Y.; Mizutani, M.; Katayama, S. High quality welding of stainless steel with 10 kW high power fibre laser. *Sci. Technol. Weld. Join.* **2009**, *14*, 288–294. [CrossRef]
28. Zhang, X.; Ashida, E.; Katayama, S.; Mizutani, M. Deep penetration welding of thick section steels with 10 kW fiber laser. *Q. J. Jpn. Weld. Soc.* **2009**, *27*, 64–68. [CrossRef]
29. Reisgen, U.; Olschok, S.; Longerich, S. Laser beam welding in vacuum—A comparison with electron beam welding. *Weld. Cut.* **2010**, *9*, 224–230.
30. Reisgen, U.; Olschok, S.; Jakobs, S. Laser beam welding in vacuum of thick plate structural steel. In Proceedings of the 32th ICALEO, Orlando, FL, USA, 6–10 October 2013; pp. 341–360.
31. Reisgen, U.; Olschok, S.; Jakobs, S.; Turner, C. Laser beam welding under vacuum of high grade materials. *Weld. World* **2016**, *60*, 1–11. [CrossRef]
32. Reisgen, U.; Olschok, S.; Jakobs, S.; Turner, C. Sound Welding of Copper: Laser Beam Welding in Vacuum. *Phys. Procedia* **2016**, *83*, 447–454. [CrossRef]
33. Börner, C.; Dilger, K.; Rominger, V.; Harrer, T.; Krüssel, T.; Löwer, T. Influence of ambient pressure on spattering and weld seam quality in laser beam welding with the solid-state laser. In Proceedings of the 30th ICALEO, Orlando, FL, USA, 23–27 October 2011; pp. 23–27.
34. Börner, C.; Krussel, T.; Dilger, K. Process characteristics of laser beam welding at reduced ambient pressure. In Proceedings of the SPIE LASE, San Francisco, CA, USA, 2–7 February 2013; p. 86030M.
35. Cai, C.; Peng, G.C.; Li, L.Q.; Chen, Y.B.; Qiao, L. Comparative study on laser welding characteristics of aluminium alloy under atmospheric and subatmospheric pressures. *Sci. Technol. Weld. Join.* **2014**, *19*, 547–553. [CrossRef]
36. Jiang, M.; Tao, W.; Wang, S.; Li, L.; Chen, Y. Effect of ambient pressure on interaction between laser radiation and plasma plume in fiber laser welding. *Vacuum* **2017**, *138*, 70–79. [CrossRef]
37. Sokolov, M.; Salminen, A.; Katayama, S.; Kawahito, Y. Reduced Pressure Laser Welding of Thick Section Structural Steel. *J. Mater. Process. Technol.* **2015**, *219*, 278–285. [CrossRef]
38. Luo, Y.; Tang, X.; Lu, F.; Chen, Q. Effect of subatmospheric pressure on plasma plume in fiber laser welding. *J. Mater. Process. Technol.* **2015**, *215*, 219–224. [CrossRef]

39. Luo, Y.; Tang, X.; Deng, S.; Lu, F.; Chen, Q.; Cui, H. Dynamic coupling between molten pool and metallic vapor ejection for fiber laser welding under subatmospheric pressure. *J. Mater. Process. Technol.* **2016**, *229*, 431–438. [CrossRef]

40. Engelhardt, T.; Heider, A.; Weber, R.; Graf, T. Time-resolved X-ray analysis of the keyhole behavior during laser welding of steel and aluminum at reduced ambient pressure. In Proceedings of the 34th ICALEO, Atlanta, GA, USA, 18–22 October 2015; pp. 250–256.

41. Elmer, J.W.; Vaja, J.; Carlton, H.D. The effect of reduced pressure on laser keyhole weld porosity and weld geometry in commercially pure titanium and nickel. *Weld. J.* **2016**, *95*, 419S–430S.

42. Heinrici, A.; Bjelajac, G.; Jonkers, J.; Jakobs, S.; Olscho, S.; Reisgen, U. Vacuum fiber-fiber coupler. In Proceedings of the SPIE LASE, San Francisco, CA, USA, 28 January–2 February 2017; p. 100970F.

43. Katayama, S.; Ido, R.; Nishimoto, K.; Mizutani, M.; Kawahito, Y. Full penetration welding of thick high tensile strength steel plate with high power disk laser in low vacuum. *Q. J. Jpn. Weld. Soc.* **2015**, *33*, 262–270. [CrossRef]

44. Youhei, A.; Yousuke, K.; Hiroshi, N.; Koji, N.; Masami, M.; Seiji, K. Effect of reduced pressure atmosphere on weld geometry in partial penetration laser welding of stainless steel and aluminium alloy with high power and high brightness laser. *Sci. Technol. Weld. Join.* **2014**, *19*, 324–332. [CrossRef]

45. Focus. Available online: http://www.focus-e-welding.de/Laser-Beam.html (accessed on 10 April 2017).

46. Welding and Joining Institute. Available online: http://www.isf.rwth-aachen.de/cms/isf/Forschung/Forschungsbereiche/Strahlschweissen/~ldah/LaserstLaserstrahlsch-unter-Vakuum/lidx/1/ (accessed on 10 April 2017).

47. Ptr-ebeam. Available online: http://www.ptr-ebeam.com/en/ptr-machines/2016-04-15-06-51-39/lasvac-pl-01.html (accessed on 10 April 2017).

48. Jakobs, S.; Reisgen, U. Laser Beam Welding under reduced pressure—Range of possible applications for thick-plates. *Stahlbau* **2015**, *84*, 635–642. [CrossRef]

49. Abe, Y.; Mizutani, M.; Kawahito, Y.; Katayama, S. Deep penetration welding with high power laser under vacuum. *Trans. JWRI* **2011**, *103*, 15–19.

50. Reisgen, U.; Olschok, S.; Turner, C. Welding of thick plate copper with laser beam welding under vacuum. *J. Laser Appl.* **2017**, *29*, 022402. [CrossRef]

51. Kim, K.R.; Farson, D.F. CO2 laser—Plume interaction in materials processing. *J. Appl. Phys.* **2001**, *89*, 681–688. [CrossRef]

52. Beck, M.; Berger, P.; Hugel, H. The effect of plasma formation on beam focusing in deep penetration welding with $CO_2$ lasers. *J. Phys. D Appl. Phys.* **1995**, *28*, 2430. [CrossRef]

53. Shcheglov, P. *Study of Vapour-Plasma Plume during High Power Fiber Laser Beam Influence on Metals*; Bundesanstalt für Materialforschung und-Prüfung (BAM): Berlin, Germany, 2012.

54. Kawahito, Y.; Kinoshita, K.; Matsumoto, N.; Mizutani, M.; Katayama, S. Effect of weakly ionised plasma on penetration of stainless steel weld produced with ultra high power density fibre laser. *Sci. Technol. Weld. Join.* **2008**, *13*, 749–753. [CrossRef]

55. Kawahito, Y.; Kinoshita, K.; Matsumoto, N.; Katayama, S. Visualization of refraction and attenuation of near-infrared laser beam due to laser-induced plume. *J. Laser Appl.* **2009**, *21*, 96–101. [CrossRef]

56. Shcheglov, P.Y.; Uspenskiy, S.A.; Gumenyuk, A.V.; Petrovskiy, V.N.; Rethmeier, M.; Yermachenko, V.M. Plume attenuation of laser radiation during high power fiber laser welding. *Laser Phys. Lett.* **2011**, *8*, 475–480. [CrossRef]

57. Pang, S.; Hirano, K.; Fabbro, R.; Jiang, T. Explanation of penetration depth variation during laser welding under variable ambient pressure. *J. Laser Appl.* **2015**, *27*, 022007. [CrossRef]

58. Pang, S.; Chen, X.; Zhou, J.; Shao, X.; Wang, C. 3D transient multiphase model for keyhole, vapor plume, and weld pool dynamics in laser welding including the ambient pressure effect. *Opt. Laser Eng.* **2015**, *74*, 47–58. [CrossRef]

59. Reisgen, U.; Olschok, S.; Jakobs, S.; Mücke, M. Welding with the laser beam in vacuum. *Laser Tech. J.* **2015**, *12*, 42–46. [CrossRef]

60. Welding and Joining Institute. Available online: http://www.isf.rwth-aachen.de/ (accessed on 10 April 2017).

61. Pro-beam. Available online: http://www.pro-beam.com/en/capabilities/fabrication/laser-systems/?no_cache=1&sword_list%5B0%5D=vacuum (accessed on 10 April 2017).

62. Laser Community. Available online: http://www.laser-community.com/en/low-pressure-solid-state-laser-welding-for-power-train-by-pro-beam-trumpf-ifs/ (accessed on 10 April 2017).
63. Weglowski, M.S.; Blacha, S.; Phillips, A. Electron beam welding—Techniques and trends—Review. *Vacuum* **2016**, *130*, 72–92. [CrossRef]
64. Letyagin, I.Y.; Trushnikov, D.N.; Belenkiy, V.Y. Benefits and Prospects of Laser Welding Application in Vacuum. *KnE Mater. Sci.* **2016**, *1*, 90–94. [CrossRef]
65. Lawler, S.; Clark, D.; Punshon, C.; Bagshaw, N.; Disney, C.; Powers, J. Local vacuum electron beam welding for pressure vessel applications. *Ironmak. Steelmak.* **2015**, *42*, 722–726. [CrossRef]
66. Schneider, A.; Gumenyuk, A.; Rethmeier, M. Mobile vacuum in pocket format. *Laser Tech. J.* **2015**, *12*, 43–46. [CrossRef]
67. Schneider, A.; Gumenyuk, A.; Rethmeier, M. Mobile vacuum device for laser beam welding of thick materials. In Proceedings of the 3rd International Conference in Africa and Asia Welding and Failure Analysis of Engineering Materials, Luxor, Egypt, 2–5 November 2015.
68. Reisgen, U.; Olschok, S.; Holtum, N.; Jakobs, S. Laser beam welding in mobile vacuum. In Proceedings of the Lasers in Manufacturing (LIM 2017), Munich, Germany, 26–29 January 2017.

*applied*
*sciences*

MDPI

*Article*

# High Power Fiber Laser Welding of Single Sided T-Joint on Shipbuilding Steel with Different Processing Setups

Anna Unt *, Ilkka Poutiainen, Stefan Grünenwald, Mikhail Sokolov and Antti Salminen

Laboratory of Laser Materials Processing, Lappeenranta University of Technology, Skinnarilankatu 34, Lappeenranta 53850, Finland; Ilkka.Poutiainen@lut.fi (I.P.); Gruenenwald@gmx.net (S.G.); mikhail.sokolov@gef.fi (M.S.); antti.salminen@lut.fi (A.S.)
* Correspondence: anna.unt@student.lut.fi; Tel.: +358-40-668-8343

Received: 2 November 2017; Accepted: 4 December 2017; Published: 8 December 2017

**Abstract:** Laser welding of thick plates in production environments is one of the main applications of high power lasers; however, the process has certain limitations. The small spot size of the focused beam produces welds with high depth-to-width aspect ratio but at times fails to provide sufficient reinforcement in certain applications because of poor gap bridging ability. The results of welding shipbuilding steel AH36 with thickness of 8 mm as a single-sided T-joint using a 10 kW fiber laser are presented and discussed in this research paper. Three optical setups with process fibers of 200 μm, 300 μm and 600 μm core diameters were used to study the possibilities and limitations set by the beam delivery system. The main parameters studied were beam inclination angle, beam offset from the joint plane and focal point position. Full penetration joints were produced and the geometry of the welds was examined. It was found that process fibers with smaller core diameter produce deeper penetration but suffer from sensitivity to beam positioning deviation. Larger fibers are less sensitive and produce wider welds but have, in turn, lower penetration at equivalent power levels.

**Keywords:** shipbuilding steel; fiber laser; laser keyhole welding; T-joint; fillet joint

---

## 1. Introduction

Laser welding with multi-kilowatt fiber lasers is fast becoming a highly advantageous joining technology in manufacturing industries such as shipbuilding, where it saves production time and cost compared to conventional arc based welding processes [1]. The growing acceptance and adoption of laser technology can be seen, for example, in sales statistics, which show an annual growth rate of over 10% for the last few years [2,3]. Modern high power fiber lasers are low maintenance, easy to integrate with production robots, and produce welds with deep penetration and low heat input at high throughput rates [4,5]. High beam quality at high power levels and the decreasing price per kilowatt of laser power are enabling previous limitations to be overcome and opening up new possibilities, especially in keyhole welding. The main limiting factor hindering more extensive utilization of laser welding is its demand for high accuracy in joint fit-up tolerances. A common way to compensate gap fluctuations and ensure welds of acceptable quality is to add an arc process working in synergy with the laser. While such hybrid laser-arc welding allows control of weld bead formation through adjustment of the arc parameters and extends the gap bridging ability of the welding system, it increases process complexity and production costs.

One of the most important characteristics of a laser welding system is the beam quality that the lasers deliver. The beam quality affects the power density of the beam, which has a direct effect on the penetration depth and geometry of the weld [6,7]. High power density of the beam means deeper penetration for the same level of power and welding speed. Single-sided welding of T-joints with fiber lasers is not an entirely novel concept, and its applicability and key parameters have been studied, for example, for aluminum welding in the aircraft industry [4]. The high depth-to-width aspect ratio of typical fiber laser welds can be a drawback in medium and thick section welding of T- and fillet joints. Space and maneuverability restrictions on the welding head can cause the laser beam to cross the joint plane at a certain angle, and a narrow melt pool may easily partially miss the joint plane and produce incomplete fusion. In addition, T-joint welds can be several meters long and heat-induced distortions during welding can thus cause variations in joint fit-up regardless of the accuracy of the original setup.

Classification societies such as DNV (Det Norske Veritas) and IIW (International Institute of Welding) suggest avoiding fillet welds in parts of a construction that are subjected to fatigue, because partial penetration creates a possible crack initiation point at the root of the weld [8,9]. Nevertheless, more than 80% of welded joints are fillet welds, because one of the plates serves as backing during the welding process and less post-welding correction is required. Figure 1 illustrates the principal differences in the geometry and location of the stress concentration of arc, laser and hybrid welded joints.

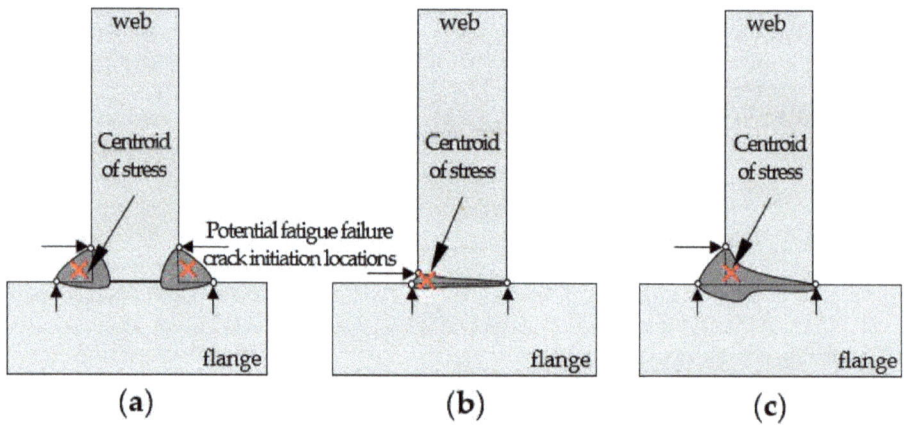

**Figure 1.** Comparison of weld joint geometries produced with: (**a**) arc welding; (**b**) autogenous laser welding; and (**c**) laser-arc hybrid welding, and stress concentration locations.

The key problem with T-joints is ensuring fusion throughout the whole joint plane, preferably with single-sided welding. In autogenous laser welding, avoidance of the possible occurrence of underfill or undercut due to an absence of filler material is important, because throat thickness and shape of the weld toe influence the fatigue performance of the joint. In thin materials and flat assembly joints, this problem can be addressed by increasing weld width through wider distribution of the beam energy. Approaches used include but are not limited to: manipulation of the focal point position [10], and usage of dual focal point setups [11–13] and beam oscillation techniques [14–17]. Scanning is also beneficial for bringing more heat into the material, which decreases the cooling rate and keeps the hardness of the weld at acceptable levels. Unfortunately, this procedure is not applicable in thick section welding, where the typical weld length is several meters, because scanning mirrors are unable to handle the power levels needed [14,18]. Large components such as scanning optics also limit the degree of freedom and flexibility of the welding process as regards positioning.

Determining the operational window for a good quality weld accounts for three main process parameters: laser power, welding speed and focal point position. These three easily adjustable parameters affect the power density at the top of workpiece, and therefore melt flow and distribution of the energy inside the keyhole, which have a major influence on the geometry of the weld. A certain threshold value of power density must be reached in order to be able to form the keyhole [19].

The threshold is typically defined as $10^6$ W/cm$^2$, or more commonly $10^3$ W/mm$^2$, with typical dimensions of laser beam focal point. The power density is calculated as follows:

$$E = \frac{P}{\pi r^2},$$ (1)

where $P$ is laser power and $r$ radius of the beam on the surface of the workpiece. The above-mentioned power density threshold is valid for $CO_2$ laser welding, whereas when using solid state lasers with wavelengths around 1000 nm the threshold is lower [20,21]. The lower threshold is a result of higher absorption of shorter wavelengths, and it gives extra flexibility to parameter selection and greater freedom to tailor parameters to specific applications, for example, with static or dynamic beam formation.

Suder and Williams et al. [22] developed the concept of Specific Point Energy ($E_{SP}$). In addition to energy density (power density × interaction time), $E_{SP}$ includes also beam diameter on the surface:

$$E_{SP} = \rho_P T_i A = P_L T_i = \frac{Pd}{v},$$ (2)

where $\rho_P$ is average power density of the beam (mW/cm$^2$), $T_i$ is interaction time (s), $A$ is area of the beam on the surface (mm$^2$), $P_L$ is laser power (W), $d$ is beam diameter on the surface (mm), and $v$ is welding speed (mm/s). Experiments performed with bead-on-plate joints have shown that power density and $E_{SP}$ control the depth of penetration and interaction time controls the bead width [22–24]. $E_{SP}$ has also been shown to be suitable for evaluating the efficiency of laser cutting [25,26].

The relationship between spot size and welding performance in steel and aluminum welding has been addressed in several studies [27–30], and it has been found that small spot size produces deeper welds yet is accompanied by defects such as undercut and porosity. A study by Vänskä [14] showed that in some cases the keyhole welding mode changes between selected parameter values, resulting in different weld cross section shape in butt joint welding of stainless steel with a disk laser. Lap and butt joints produced with $CO_2$ and solid-state laser sources have been characterized and compared [20]; for example, Kawahito et al. [28] addressed the effect of focal spot size on weld defects and showed that, of the four sizes studied, welds with highest quality were obtained using the two larger spot sizes.

The effect of focal point diameter on beam intensity on surface is much stronger than the effect via laser power. A simple way to manipulate the focal point diameter is to change the focal length of the focusing lens. A problem with this approach is that the focusing angle, i.e., the angle at which the beam enters the keyhole, changes as the focal length is changed, and the effect of focal point diameter is influenced by the effect of focusing angle and changes in the shielding gas arrangement. Change in the beam feeding fiber can only increase the diameter of the raw beam and the beam parameter product. This paper addresses the issue of weld quality of high power fiber laser welded T-joints of shipbuilding steel AH36 by comparing welds produced with three optical set-ups having beam transfer fibers of different diameters.

## 2. Materials and Methods

Shipbuilding steel AH36 is commonly used for shipbuilding and offshore structures. Hot rolled steel plates of AH36 have excellent weldability with a CEV (carbon equivalent value) of 0.248, calculated based on the chemical composition presented in Table 1. The yield strength of AH36 is 355 MPa.

Table 1. Chemical composition of AH36 steel (wt %).

| Material | C | Si | Mn | P | S | Cr | Mo | Ni | Cu | Al | V |
|---|---|---|---|---|---|---|---|---|---|---|---|
| AH36 | 0.111 | 0.149 | 0.711 | 0.035 | 0.150 | 0.051 | 0.01 | 0.041 | 0.031 | 0.030 | 0.008 |

Test specimens (100 mm × 350 mm × 8 mm) were cut with a $CO_2$ laser using oxygen-assisted cutting. The edges were grid blasted with aluminum oxide and cleaned with acetone to remove possible contaminants. The plates were tack welded from the root side from the ends and the middle using gas metal arc welding. The workpiece was fixed in the flat (1F) position using stiff fixtures to avoid heat-induced air gap fluctuations during the welding. Single-sided welds with a length of 165 mm were performed. The experimental setup is shown in Figure 2.

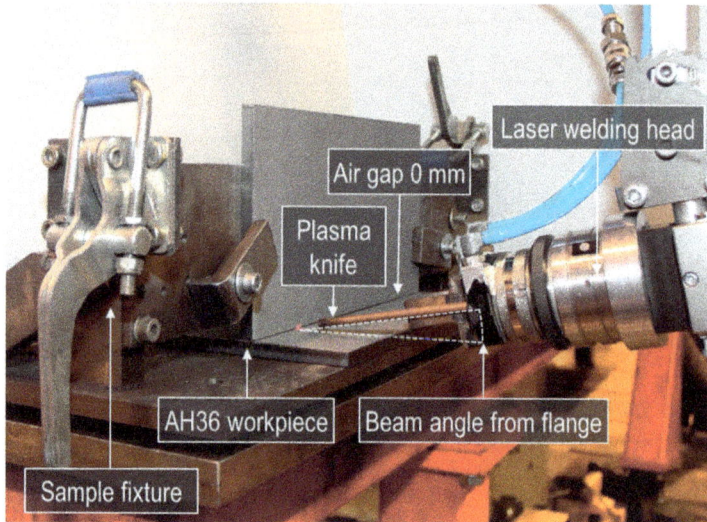

**Figure 2.** Experimental setup for laser welding of T-joints. Beam is positioned at the joint plane (offset 0 mm).

All welding experiments were made with a continuous wave fiber laser IPG YLS-10000 having wavelength of 1070 nm and a top-hat focused beam profile. A Kugler LK190 mirror optics laser welding head was used. An air knife protected the focusing system from contamination with fumes and spatter, and no additional shielding gas was used. When comparing a T-joint with a butt joint, a further parameter, called offset, describing the distance of the beam center from the joint at the flange front edge, must be added. This parameter has a crucial effect on weld quality and must be considered for thick section joints of the type studied in this work. The experimental parameters of the welding process are presented in Table 2.

**Table 2.** Welding process parameters.

| Parameter | Unit | Parameter Range |
|---|---|---|
| Fiber diameter | [μm] | 200; 300; 600 |
| Laser power, $P_L$ | [kW] | 6.0; 8.0; 10.0 |
| Welding speed, $v_w$ | [m/min] | 0.75; 1.0; 1.25; 1.5; 1.75 |
| Focal point position, $F_{PP}$ | [mm] | −2.0; −4.0; −6.0 |
| Beam angle from flange α | [°] | 6; 10; 15 |
| Beam offset from flange | [mm] | 0.5; 1.0; 1.2; 1.5; 2.0 |

The beam was delivered by a system consisting of a beam transfer fiber (with core diameters of either 200, 300 or 600 μm), 120 mm collimating length optics, and a 300 mm focal length mirror. The properties of the laser beam emitted from each transport fiber were measured using a laser beam analyzer from Primes GmbH and are shown in Table 3.

**Table 3.** Beam properties.

| Delivery Fiber Diameter (µm) | 200 | 300 | 600 |
|---|---|---|---|
| Beam profile | | | |
| Nominal beam waist (mm) | 0.50 | 0.75 | 1.50 |
| Measured beam waist (86% pts) (mm) | 0.710 | 0.882 | 1.460 |
| BPP (mm·mrad) | 9.079 | 12.000 | 23.800 |
| Rayleigh length (mm) | 13.86 | 16.18 | 22.38 |
| $P_L$ at workpiece (kW) | 6.0 | 6.0 | 6.0 |
| Beam area at surface (mm$^2$) | 0.396 | 0.611 | 1.674 |

The bead surface and root of each weld were visually evaluated based on standard EN ISO 13919-1, which classifies welds into three quality levels based on the type and severity of the imperfections that are present. The categories from best to worst are: B, stringent; C, intermediate; and D, moderate [31]. Metallographic preparation of the samples was carried out according to SFS-EN ISO 17639 [32]. The welds were transversely sectioned at the middle of the joint length, and polished and etched using a 2% Nital solution. Macrographs of the weld cross sections were taken for inspection of penetration, defects, and dimensions and shape of the fusion zone and HAZ (heat affected zone).

## 3. Results

### 3.1. Effect of Beam Inclination Angle α

To study the effect of beam inclination angle α on the morphology/geometry of the weld, all other process parameters were kept constant. The beam was positioned 0.5 mm above the joint plane on the flange and focused 2 mm below the surface of the material. Focal point diameters on the surface were 0.82 mm, 1.00 mm and 1.61 mm for the 200 µm, 300 µm and 600 µm process fibers, respectively. The cross sections of the welds are shown in Figure 3.

It can be seen in Figure 3 that the penetration depth (measured from top of the weld) and the length of the joint fusion along the intersection decreased as the inclination angle increased. It can also be noticed that the penetration depth and area of the fusion zone correlate with the energy density of the beam. A full penetration weld was obtained only with the 200 µm process fiber and 6° inclination angle. All of the joints followed the axis of beam propagation. Averaged dimensions from three welds produced with each inclination angle are shown in Table 4.

**Table 4.** Effect of fiber diameter on the weld dimensions.

| Fiber Diameter [µm] | Penetration Depth [mm] | Bead Width [mm] | Fusion Zone [mm$^2$] | HAZ Area [mm$^2$] | Depth to Width Ratio | Max Hardness HV5 [FZ [1]/HAZ] |
|---|---|---|---|---|---|---|
| 200 | 8.7 | 2.3 | 13.2 | 6.5 | 4.0 | 386/373 |
| 300 | 7.8 | 2.4 | 12.6 | 6.0 | 3.3 | 392/359 |
| 600 | 5.6 | 2.7 | 12.0 | 5.5 | 2.0 | 393/365 |

[1] FZ = fusion zone.

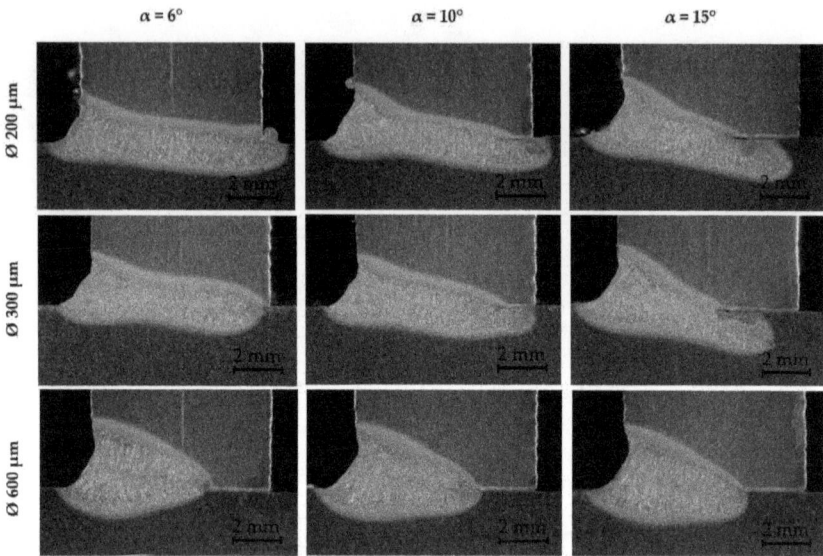

**Figure 3.** Macrographs of weld samples at different beam inclination angles: AH36, $t = 8$ mm, $P_L = 6$ kW, $v_w = 1.25$ m/min, beam offset from flange 0.5 mm, $F_{PP} = -2$ mm.

### 3.2. Effect of Beam Offset from the Flange

The effect of the beam offset from the flange was studied only with 300 μm and 600 μm process fibers. Based on the very narrow bead width produced earlier with the 200 μm process fiber and an assumption that the weld width largely determines the tolerance limits, the 200 μm process fiber was not included in the experiments. The laser power needed for full penetration was first calculated using the Power Factor Model developed by Suder et al. [23] and subsequently determined experimentally. The calculated values exceeded the real power requirement by at least 30% (less than 8 kW vs. 9.7 kW, $F_{PP}$ −2 mm, beam Ø on surface 1.0 mm; 10 kW vs. 13 kW, $F_{PP}$ −2 mm, beam Ø on surface 1.6 mm). Macrographs showing beam offsets from 0.5 mm to 2.0 mm are presented in Figure 4.

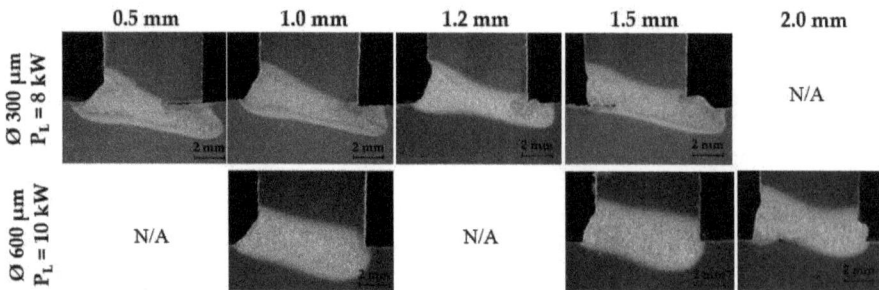

**Figure 4.** Macrographs of weld cross-sections when beam position from the flange was varied: material = AH36, $t = 8$ mm, $v_w = 1.25$ m/min, $F_{PP} = -2$ mm, $\alpha = 15°$.

As shown in Figure 4, despite similar width of weld bead, there was a significant difference in the geometry of the fusion area. Welds produced with the 300 μm process fiber had a deep and narrow profile with a slightly wider top typical of high power laser welds. The setup with the 600 μm process fiber produced welds that were wide throughout the whole fusion area, resulting in a more acceptable weld profile for a T-joint.

## 3.3. Effect of Focal Point Position

Negative defocusing was used to study whether decrease in beam density at the workpiece surface has a favorable effect on the formation of the weld bead. The influence of focal point position on the weld profile was investigated by changing the defocusing distance in steps of 2 mm. Beam offset from the flange was selected as 1 mm and 1.5 mm (for the 300 μm and 600 μm setups, respectively) to increase the likelihood of full penetration. The focal point was moved along the beam propagation direction inside the material in 2 mm steps. Figure 5 shows cross-sectional macrographs of the welds produced.

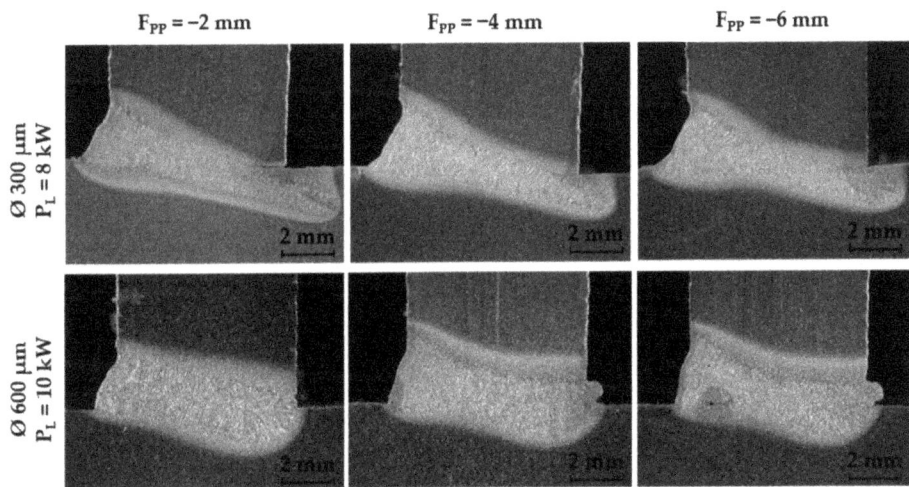

**Figure 5.** Weld profiles at various focal point positions: AH36, $t$ = 8 mm, $v_w$ = 1.25 m/min, beam offset from flange 1.0 mm (up) and 1.5 mm (down), $\alpha$ = 15°.

It can be seen from the images presented in Figure 5 that decreasing the focal point position leads to a slight decrease in penetration. $F_{PP}$ −4 mm resulted in full penetration in the set-up with the 600 μm process fiber, while none of the welds produced with the 300 μm fiber had complete penetration at the weld root.

## 4. Discussion

### 4.1. Geometry of the Welds

The purpose of this study was to investigate geometrical differences in welds produced with three different beam delivery fibers and to determine the effect of process parameters on T-joint welds. Thirty welds were produced, evaluated, and their cross-sections analyzed. The acceptance criteria in visual inspection were smooth and plain face and root sides of the weld seam, lack of spatter, cracks or other defects listed in the Standard EN ISO 13919-1, and full visible penetration on the root side. Cracks or porosity were not present, qualifying the welds for class C of ISO 13919-1. Obvious undercut and lack of fusion produced by an inappropriately positioned beam or a lack of laser power were causes of rejection.

The width of the weld fusion zone determines the largest acceptable inclination angle for producing full penetration at a given web thickness. It can be seen in Figure 3 that compared to the other set-ups studied, the process fiber with a core diameter 600 µm was least sensitive to increase of α. However, at given thickness of 8 mm, α = 6° produced the largest weld throat in all set-ups, regardless of the diameter of the beam, since the keyhole formed strictly along the axis of beam propagation. Figure 6 illustrates the effect of beam inclination angle on melt distribution relative to the middle axis of the joint.

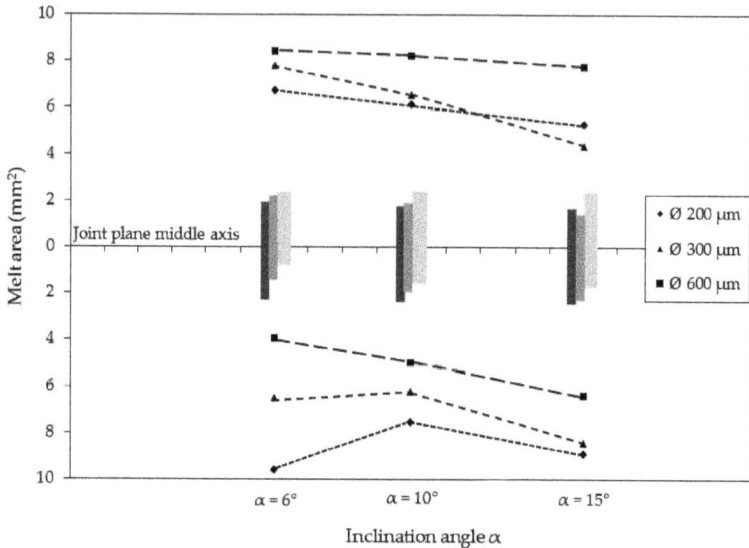

**Figure 6.** Proportions of the weld and melt area distribution differences at three inclination angles tested. AH36, t = 8 mm, $P_L$ = 6 kW, $v_w$ = 1.25 m/min, $F_{PP}$ = −2 mm.

Figure 6 shows the size of the melt area above and below the middle axis of the joint of the welds presented earlier in Figure 3. The bars illustrate the proportions of the melt area above (web) and below (flange) the joint plane, and the lines show the area of the melt. Process fibers with core diameters of 200 µm and 300 µm produced welds with similar properties: the melt area decreased with increase of inclination angle. The 600 µm diameter process fiber produced welds in which the melt area increased while simultaneously keeping the same proportions of the fusion zone above and below the joint axis. The fiber producing the largest focal point diameter was least sensitive to change of inclination angle.

From an engineering point of view, beam offset is an important parameter affecting the weld quality of fillet welds, where joint bridging ability and ensuring complete fusion are more important than the penetration depth itself. As can be observed in Figure 4, tolerance to beam offset is also determined by the width of the weld. An inclination angle of 15° was chosen over two smaller angles tested based on the width of the weld bead and possible accessibility restrictions of the welding head. Optimal offsets for acceptable top bead and full fusion at the root are: 1.0 mm with the 300 µm transfer fiber, and, 1.0 mm and 1.5 mm with the 600 µm transfer fiber. The diameters of the beams on the surface were 1.0 mm and 1.6 mm, respectively.

When the 300 µm fiber was used, deviation to either side from the 1.0 mm offset produced either lack of fusion at the root or undercut and lack of penetration at the face of the weld. Due to the small width of the weld/fusion zone, higher offset resulted in formation of severe undercut on the top of the weld, while part of the melt was pushed through the weld root. The transfer fiber with 600 µm core diameter had a wider positioning tolerance window because the width of the weld throughout the fusion area is also wider. All of the welds produced with the 600 µm process fiber had a smooth bead and root sides, complete penetration, and class B quality according to EN-ISO 13919-1. It seems that a change in the keyhole process, noticed earlier by [14], can be seen in the case of fillet welds in low alloyed steel. It is logical that power density has an effect on the process mechanism while still producing weld shape that is similar but wider than in the case of higher power density.

The focal point position has the same effect on T-joints as any other weld. Comparing the setups studied, an insignificant increase in bead width in correlation with an increase in beam dimension on the surface of material was noticed. $F_{PP}$ of $-2$ mm produced welds with a larger weld toe radius than $F_{PP}$ $-6$ mm. When $F_{PP}$ was positioned at $-4$ mm, that is, at half of the thickness of the web, the setup produced fully fused welds lacking porosity or other defects on both sides of the joint.

### 4.2. Application of Specific Point Energy to Welding of T-Joints

Traditionally, the depth of penetration has been characterized through the concept of heat input, also called line energy, which describes the energy available for producing the weld through the relationship of available laser power and welding speed. $E_{SP}$ also considers the diameter of the beam on the surface of the specimen and therefore provides higher accuracy in definition of the penetration depth and weld geometry [23]. Suder and Williams [22] have shown that, in the case of bead-on-plate welds, the penetration depth is defined by power density and specific point energy, and the width of the weld by interaction time, regardless of the optical setup used. T-joint fillet welds follow the same analytical model regarding the penetration depth as bead-on-plate welds; in this work, however, there was a deviation at larger beam diameters produced with the 600 μm process fiber. The relationship between $E_{SP}$ and penetration depth for all three setups is summarized in Figure 7.

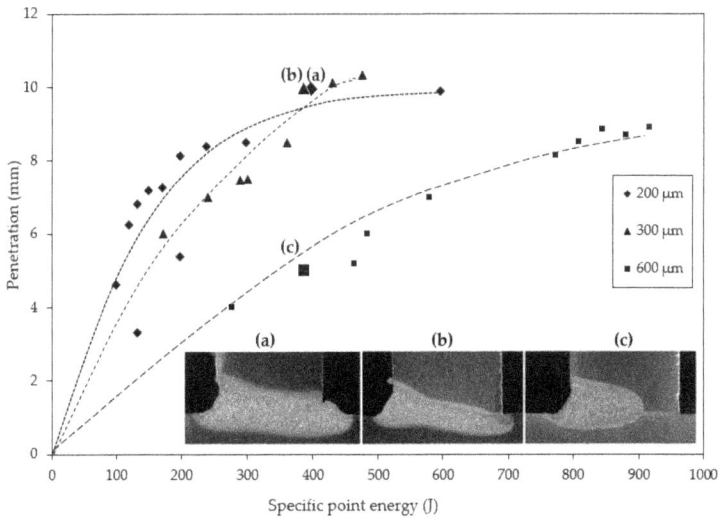

**Figure 7.** Penetration depth plotted as a function of specific point energy. $E_{SP}$ values of the three welds shown: (▲a) 396 J; (◆b) 384 J; and (■c) 386 J.

Figure 7 shows curves for penetration depth produced with each of the process fibers and macrographs of three welds obtained at similar $E_{SP}$ using different process fibers. Beam diameters on the surface of the welds shown in Figure 7 were 0.82 mm, 1.00 mm and 1.61 mm ($F_{PP}$ $-2$ mm). The following laser power and welding speed combinations were used: (a) $P_L = 6$ kW, $v_w = 0.75$ m/min; (b) $P_L = 8$ kW, $v_w = 1.25$ m/min; and (c) $P_L = 4$ kW, $v_w = 1$ m/min. Welds (a) and (b) both had penetration depths close to 10 mm, when measured from the top of the bead, thus exceeding the thickness of the material, while weld (c) had only partial penetration. At similar $E_{SP}$, the 200 μm process fiber produced the weld with the deepest penetration and largest melt area (a). Full penetration was not

*Appl. Sci.* **2017**, *7*, 1276

obtained with the 300 μm process fiber despite sufficient penetration depth, and the weld produced with the 600 μm fiber was noticeably shallower.

It is known that the diameter of the beam governs the dimensions of the keyhole, which in turn define the depth and width of the weld. As expected, beams with smaller diameters produced deeper penetrations at any given $E_{SP}$. Within each setup, it can be seen that an increase in $E_{SP}$ also increases the penetration depth, but comparison of all three setups shows that the power density of the beam has a more pronounced effect on the morphology of the weld than $E_{SP}$. A possible explanation for this might be that the heat conduction in T-joints is different than in butt joints or bead on plate joints. Distribution of energy over a larger area leads to a slight decrease in molten area and reduced penetration depth. The power density of the beam determines the depth of the weld, while the width of the weld is determined by the diameter of the focused beam on the surface of the specimen.

*4.3. Optimal Welding Conditions for T-Joint*

The effects of beam inclination angle, beam offset from the flange and the focal point position relative to the surface of the material were studied to gain insight into the applicability of each setup under industrial conditions. Table 5 summarizes the findings.

**Table 5.** Acceptable limits for beam positioning for producing full penetration.

| Parameter | 200 μm Process Fiber | 300 μm Process Fiber | 600 μm Process Fiber |
| --- | --- | --- | --- |
| $\alpha$ | 6° | 6° | 6°; 10° |
| Beam offset | 1 mm | 1 mm | 1–1.5 mm |
| $F_{pp}$ | −4 mm | −4 mm | −2–−6 mm |

The optimal parameters for all set-ups were inclination angle 6°, beam offset from the flange 1 mm, and focal point position −4 mm below the surface of the material. In all of the setups, the axis of the weld was aligned along the direction of the beam propagation. For this reason, when the beam was aimed past the root of the joint, the formed molten pool did not follow the joint plane, resulting in a lack of fusion at the back of the weld. However, the applicability of a beam inclination angle of 6° in industrial applications may be limited because of a danger of collision of the laser welding head while maneuvering in restricted space, especially in situations where the incident beam side of the flange exceeds the focal length of the laser.

From the industrial point of view, the most versatile solution for T-joints of the three process fibers tested would be the fiber with a core diameter of 600 μm. This setup produced top beads superior in quality to the two other setups studied. The welds made with the 200 μm and 300 μm process fibers were deep yet extremely narrow at the deepest section of the weld and prone to undercut at the surface. A setup with a 600 μm fiber results in a more stable process that has a greater tolerance for beam displacement and smaller probability for seam imperfections.

**5. Conclusions**

The present work reported welding of single-sided T-joints of 8 mm thick AH36 shipbuilding steel with three optical setups using process fibers with core diameters of 200 μm, 300 μm and 600 μm. The current study found that:

(1) Full fusion in one welding pass was produced with all three process fibers studied.
(2) Penetration depth and width of the weld both primarily depend on the beam diameter. The parameter with the greatest influence on the depth of the weld is the power density of the beam, while the width of the weld is determined by the diameter of the focused beam. The width of the weld bead only has a minor correlation to the diameter of the beam on the surface.

(3)  Smaller spot sizes provide an advantage in penetration depth at the same welding speed and power but are prone to producing undercuts. Due to the narrowness of the weld, the positioning of the beam has to be extremely accurate to avoid the weld missing the root of the joint.

(4)  Welds produced with 600 μm process fiber were less prone to undercut formation and had more favorable shape of the weld toe than welds produced with 200 μm and 300 μm process fibers.

(5)  Process fiber with core diameter 600 μm produced welds with the highest quality and was least sensitive to changes in beam positioning.

(6)  Using beam delivery fibers with larger core diameters has a favorable effect on achieving full fusion in T-joints. Reduced energy density on surface increases the width of the weld throughout the penetration and produces smoother junctions of weld bead and base material.

**Acknowledgments:** Authors gratefully acknowledge Pertti Kokko for assistance with the experiments and Antti Heikkinen for the help with the metallography. Authors would like to thank the project PAMOWE of Academy of Finland for financial support.

**Author Contributions:** Anna Unt, Ilkka Poutiainen and Antti Salminen conceived and designed the experiments; Anna Unt and Ilkka Poutiainen performed the experiments; Anna Unt, Stefan Grünenwald, Mikhail Sokolov and Antti Salminen analyzed the data; Antti Salminen and Ilkka Poutiainen contributed materials/analysis tools; and Anna Unt and Antti Salminen wrote the paper.

**Conflicts of Interest:** The authors declare no conflicts of interest.

## References

1. Grupp, M.; Klinker, K.; Cattaneo, S. Welding of high thicknesses using a fibre optic laser up to 30 kW. *Weld. Int.* **2013**, *27*, 109–112. [CrossRef]
2. Belforte, D. Laser Market Results Confound the Experts. Available online: http://www.industrial-lasers.com/articles/2017/01/laser-market-results-confound-the-experts.html (accessed on 16 August 2017).
3. Thoss, A.F. Four laser companies to exceed $1 billion revenue in 2016. *Adv. Opt. Technol.* **2017**, *6*, 13–16. [CrossRef]
4. Enz, J.; Khomenko, V.; Riekehr, S.; Ventzke, V.; Huber, N.; Kashaev, N. Single-sided laser beam welding of a dissimilar AA2024–AA7050 T-joint. *Mater. Des.* **2015**, *76*, 110–116. [CrossRef]
5. Liu, S.; Mi, G.; Yan, F.; Wang, C.; Jiang, P. Correlation of high power laser welding parameters with real weld geometry and microstructure. *Opt. Laser Technol.* **2017**, *94*, 59–67. [CrossRef]
6. Sokolov, M.; Salminen, A. Improving laser beam welding efficiency. *Engineering* **2014**, *6*, 559–571. [CrossRef]
7. Kuryntsev, S.V.; Gilmutdinov, A.K. Welding of stainless steel using defocused laser beam. *J. Constr. Steel Res.* **2015**, *114*, 305–313. [CrossRef]
8. Hobbacher, A. *Recommendations for Fatigue Design of Welded Joints and Components*; Springer: Berlin, Germany, 2015.
9. Det Norske Veritas. *Fatigue Design of Offshore steel Structures, Recommended Practice*; DNV-RP-C203; DNV GL: Oslo, Norway, 2008.
10. Matsumoto, N.; Kawahito, Y.; Nishimoto, K.; Katayama, S. Effects of laser focusing properties on weldability in high-power fiber laser welding of thick high-strength steel plate. *J. Laser Appl.* **2017**, *29*, 012003. [CrossRef]
11. Grajcar, A.; Morawiec, M.; Różański, M.; Stano, S. Twin-spot laser welding of advanced high-strength multiphase microstructure steel. *Opt. Laser Technol.* **2017**, *92*, 52–61. [CrossRef]
12. Morawiec, M.; Różański, M.; Grajcar, A.; Stano, S. Effect of dual beam laser welding on microstructure–property relationships of hot-rolled complex phase steel sheets. *Arch. Civ. Mech. Eng.* **2017**, *17*, 145–153. [CrossRef]
13. Shen, J.; Li, B.; Hu, S.; Zhang, H.; Bu, X. Comparison of single-beam and dual-beam laser welding of Ti–22Al–25Nb/TA15 dissimilar titanium alloys. *Opt. Laser Technol.* **2017**, *93*, 118–126. [CrossRef]
14. Vänskä, M. Defining the Keyhole Modes—The Effects on the Weld Geometry and the Molten Pool Behaviour in High Power Laser Welding of Stainless Steels. Ph.D. Thesis, Lappeenranta University of Technology, Lappeenranta, Finland, 2014.
15. Müller, A.; Goecke, S.F.; Sievi, P.; Albert, F.; Rethmeier, M. Laser beam oscillation strategies for fillet welds in lap joints. *Phys. Proc.* **2014**, *56*, 458–466. [CrossRef]
16. Wang, L.; Gao, M.; Zhang, C.; Zeng, X. Effect of beam oscillating pattern on weld characterization of laser welding of AA6061-T6 aluminum alloy. *Mater. Des.* **2016**, *108*, 707–717. [CrossRef]

17. Hao, K.; Li, G.; Gao, M.; Zeng, X. Weld formation mechanism of fiber laser oscillating welding of austenitic stainless steel. *J. Mater. Process. Technol.* **2015**, *225*, 77–83. [CrossRef]
18. Zhang, M.; Chen, G.; Zhou, Y.; Liao, S. Optimization of deep penetration laser welding of thick stainless steel with a 10 kW fiber laser. *Mater. Des.* **2014**, *53*, 568–576. [CrossRef]
19. Ion, J. *Laser Processing of Engineering Materials: Principles, Procedure and Industrial Application*; Butterworth-Heinemann: Oxford, UK, 2005; p. 179. ISBN 0-7506-6079-1.
20. Zou, J.L.; He, Y.; Wu, S.K.; Huang, T.; Xiao, R.S. Experimental and theoretical characterization of deep penetration welding threshold induced by 1-μm laser. *Appl. Surf. Sci.* **2015**, *57*, 1522–1527. [CrossRef]
21. Courtois, M.; Carin, M.; Le Masson, P.; Gaied, S.; Balabane, M. A new approach to compute multi-reflections of laser beam in a keyhole for heat transfer and fluid flow modelling in laser welding. *J. Phys. D Appl. Phys.* **2013**, *46*, 505305. [CrossRef]
22. Suder, W.J.; Williams, S.W. Investigation of the effects of basic laser material interaction parameters in laser welding. *J. Laser Appl.* **2012**, *24*, 032009. [CrossRef]
23. Suder, W.J.; Williams, S. Power factor model for selection of welding parameters in CW laser welding. *Opt. Laser Technol.* **2014**, *56*, 223–229. [CrossRef]
24. Ayoola, W.A.; Suder, W.J.; Williams, S.W. Parameters controlling weld bead profile in conduction laser welding. *J. Mater. Process. Technol.* **2017**, *249*, 522–530. [CrossRef]
25. Nikhare, N.B.; Arakerimath, R.R. Parametric analysis and heat transfer enhancement of laser welding for different material. *Int. J. Eng. Manag. Res.* **2015**, *ICRAME-2015*, 92–96.
26. Hashemzadeh, M.; Suder, W.; Williams, S.; Powell, J.; Kaplan, A.F.H.; Voisey, K.T. The application of specific point energy analysis to laser cutting with 1 μm laser radiation. *Phys. Proc.* **2014**, *56*, 909–918. [CrossRef]
27. Verhaeghe, G. The effect of spot size and laser quality on welding performance when using high-power continuous wave solid-state lasers. In Proceedings of the ICALEO'2005 Conference, Miami, FL, USA, 31 October–3 November 2005; pp. 264–271.
28. Kawahito, Y.; Mizutani, M.; Katayama, S. Investigation of high-power fiber laser welding phenomena of stainless steel. *Trans. JWRI* **2007**, *36*, 11–15.
29. Bhargava, P.; Paul, C.P.; Mundra, G.; Premsingh, C.H.; Mishra, S.K.; Nagpure, D.; Kumar, A.; Kukreja, L.M. Study on weld bead surface profile and angular distortion in 6 mm thick butt weld joints of SS304 using fiber laser. *Opt. Laser Eng.* **2014**, *53*, 152–157. [CrossRef]
30. Katayama, S.; Kawahito, Y.; Mizutani, M. Elucidation of laser welding phenomena and factors affecting weld penetration and welding defects. *Phys. Proc.* **2010**, *5*, 9–17. [CrossRef]
31. International Organization for Standardization. *EN ISO 13919-1: Welding—Electron and Laser-Beam Welded Joints—Guidance on Quality Levels for Imperfection—Part 1: Steel*; ISO: Geneva, Switzerland, 1996; 9p.
32. International Organization for Standardization. *EN ISO 17639 Destructive Tests on Welds in Metallic Materials—Macroscopic and Microscopic Examination of Welds*; ISO: Geneva, Switzerland, 2003.

*applied*
*sciences*

MDPI

*Article*

# A Comparative Study on the Laser Welding of Ti6Al4V Alloy Sheets in Flat and Horizontal Positions

**Baohua Chang [1], Zhang Yuan [1], Haitao Pu [1], Haigang Li [2], Hao Cheng [2], Dong Du [1],\*
and Jiguo Shan [1],\***

[1]  State Key Laboratory of Tribology, Department of Mechanical Engineering, Tsinghua University,
    Beijing 100084, China; bhchang@tsinghua.edu.cn (B.C.); 15201518430@163.com (Z.Y.);
    puhaitaohehe@163.com (H.P.)
[2]  Aerospace Research Institute of Materials & Processing Technology, Beijing 100076, China;
    lhg703@sina.com (H.L.); chenghao611@126.com (H.C.)
\*  Correspondence: dudong@tsinghua.edu.cn (D.D.); shanjg@tsinghua.edu.cn (J.S.);
    Tel.: +86-10-6278-1182 (D.D. & J.S.); Fax: +86-10-6277-3862 (D.D. & J.S.)

Academic Editor: Federico Pirzio
Received: 21 March 2017; Accepted: 6 April 2017; Published: 10 April 2017

**Abstract:** Laser welding has been increasingly utilized to manufacture a variety of components thanks to its high quality and speed. For components with complex shapes, the welding position needs be continuously adjusted during laser welding, which makes it necessary to know the effects of the welding position on the quality of the laser welds. In this paper, the weld quality under two (flat and horizontal) welding positions were studied comparatively in the laser welding of Ti6Al4V titanium alloy, in terms of weld profiles, process porosity, and static tensile strengths. Results show that the flat welding position led to better weld profiles, less process porosity than that of the horizontal welding position, which resulted from the different actions of gravity on the molten weld metals and the different escape routes for pores under different welding positions. Although undercuts showed no association with the fracture positions and tensile strengths of the welds, too much porosity in horizontal laser welds led to significant decreases in the strengths and specific elongations of welds. Higher laser powers and travel speeds were recommended, for both flat and horizontal welding positions, to reduce weld porosity and improve mechanical properties.

**Keywords:** titanium alloy; laser welding; welding position; porosity; weld profile

## 1. Introduction

Titanium alloys have been widely used in many industrial fields, such as aerospace and aircraft, because of their superior properties [1]. Meanwhile, lasers are applied to weld such titanium alloy components and achieve high quality at a high speed thanks to its high brightness and power availability [2]. For components with complex shapes, the weld tracks are generally not straight lines but complicated two-dimensional or three-dimensional curves (e.g., girth welds of pipelines), which lead to changes in the welding positions during laser welding. Such changes in welding positions may result in fluctuations in welding quality because of the different actions of gravity for various welding positions, which then necessitates the adjustment of welding parameters accordingly. As a basis to optimize the laser welding parameters for varying welding positions, it is therefore necessary to study the influence of welding position on weld quality.

Some research has already been done on the effects of welding position in the laser welding of steels. Guo et al. [3] indicated that employing the 2G (horizontal) welding position (with the laser beam perpendicular to the direction of gravity) could mitigate the welding defects of undercuts and sagging in laser welding of 13-mm-thickhigh strength steel plates. Such defects occurred commonly when using the 1G (flat) welding position (with the laser beam in the same direction with gravity). Such an improvement in quality was attributed to a more balanced state for the weld pool between the surface tension, recoil pressure, and gravity. Shen et al. [4] compared the process window and porosity distribution when laser welding 10-mm-thick 30CrMnSiA ultrahigh strength steel plates using flat and horizontal positions, and found that higher line energies were required to reach the same penetration depth in the horizontal position than that for the flat position. In addition, the pores in the horizontal position laser welds were located in the upper part of the weld, while those in the flat position were in the weld center. Sohail et al. [5] studied the laser welding of 20-mm-thick low carbon steel plates in eight different welding positions, and found that the welding position had little influence on the weld bead shape and the fluid flow features of weld pools, but could lead to different levels and positions of porosity.

Besides laser welding, the effects of the welding position have also been studied for other welding processes, such as gas metal arc welding (GMAW), hybrid laser-arc welding (HLAW), and electron beam (EB) welding. Kumar and Debroy [6] numerically investigated the fluid flow during gas-metal-arc fillet welding for different various joint configurations (L and V shapes) and welding positions (tilting angles), and revealed that workpiece orientation and welding configuration could affect the free surface profile of the weld pool significantly, which might in turn affect the strength of the welds. Cho et al. [7] numerically studied the molten pool behaviors in gas metal arc welding (GMAW) of a 10-mm-thick V-groove steel plate with various welding positions (flat, overhead, and vertical). Different trends were found for humping, overflow, and lack of penetration. Lin et al. [8–10] investigated the molten pool behavior for all-position narrow gap GMAW of 25-mm-thick carbon steel plates and found that the molten pool surfaces had different shapes when welding in flat, vertical down, and overhead welding positions due to the different actions of gravity. Chen et al. [11] studied the effects of welding position on the droplet transfer behavior during hybrid $CO_2$ laser-MAG welding of 16-mm-thick steel and indicated that the gravity, in combination with the electromagnetic force, would cause great differences in impacting positions, modes, dimensions, and frequencies of the droplets for different welding positions (flat, horizontal, and vertical). The study by Koga et al. [12] on the all-positional electron beam welding of 19-mm-thick pipeline steel plates indicated that the optimal welding parameters were different for different welding positions, and therefore need be adjusted appropriately during a girth welding.

Existing research work on positional welding has focused mainly on thick steels used for construction of pipelines, and only a few of them are concerned with laser beam welding. Within those references, the laser welding of titanium alloy sheets under different welding positions has not been reported so far. Therefore, this research sets out to study the influence of two welding positions (flat and horizontal) on the weld quality in terms of weld profile, porosity, and strength, when laser welding Ti6Al4V titanium alloy. The results obtained can then be useful in process optimization when laser welding titanium alloy components with complex joint shapes.

## 2. Experimental Procedures

Titanium alloy (Ti6Al4V, annealed) sheets 3 mm in thickness were used in the study, with the chemical composition listed in Table 1. The sheets were cut into rectangular workpieces of approximately 150 mm × 300 mm. Argon with a purity of 99.998% was used as a shielding gas in all instances.

Table 1. Chemical composition of the titanium alloy Ti6Al4V sheets (wt %).

| Elements | Al | V | Fe | C | N | H | O | Ti |
|---|---|---|---|---|---|---|---|---|
| Contents, wt % | 5.8 | 4.0 | 0.2 | 0.05 | 0.03 | 0.011 | 0.19 | Balance |

Welding trials were performed using an IPG Photonics YLS-6000 (6 kW) Yb-fiber laser (IPG Photonics, Oxford, MS, USA) with an output wavelength of 1070 ± 10 nm. Table 2 details the collimating and focusing units used and the resulting calculated laser beam profile characteristics. For all the trials performed, the process head was mounted on a 6-axis articulated arm robot (Reis Robotics, Obernburg, Bavaria, Germany).

**Table 2.** Optic combinations used in the butt welding trials.

| Parameter | Value |
| --- | --- |
| Delivery fiber core diameter, mm | 0.2 |
| Beam parameter product, mm·mrad | 6 |
| Collimating unit focal length, mm | 100 |
| Focusing unit focal length, mm | 300 |
| Beam width, mm | 0.6 |
| Rayleigh length, mm | 0.94 |

A schematic diagram of the laser welding setup is shown in Figure 1. Three flows of shielding gas were employed in welding. The main flow was used to protect the molten weld pool, with a rate of 20 L/min. The trailing shielding flow was used to prevent the high temperature metals that just solidified from oxidation, with a rate of 70 L/min. The root shielding flow was to protect the back surface of a specimen, with a rate of 5 L/min.

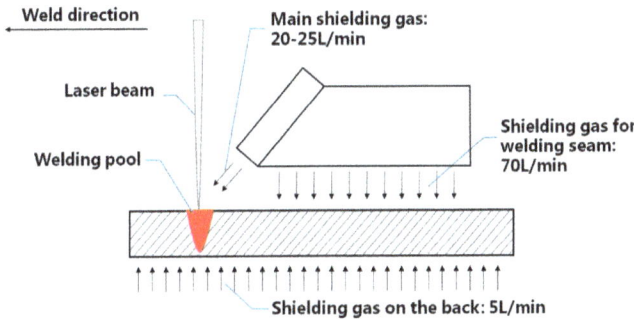

**Figure 1.** Schematic diagram of the laser welding setup.

The two welding positions studied (i.e., flat and horizontal) are shown in Figure 2. For each welding position, two sets of laser welding parameters were chosen, as listed in Table 3. In all welding trials, the focal position of the laser beam was located on the top surface of the workpiece.

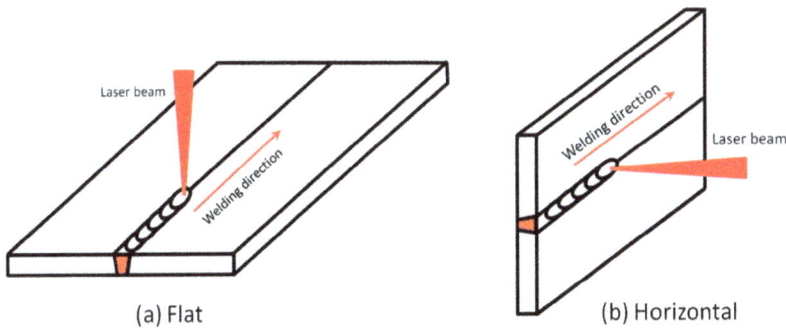

(a) Flat    (b) Horizontal

**Figure 2.** Two welding positions studied in the study.

**Table 3.** Welding parameters used in the experiments.

| Welding Positions | Butt Weld Identity | Laser Power, (kW) | Welding Speed, (mm/s) | Heat Input (J/mm) |
|---|---|---|---|---|
| Flat | F01 | 2.2 | 8 | 275 |
| | F02 | 2.5 | 20 | 125 |
| Horizontal | H01 | 2.2 | 8 | 275 |
| | H02 | 2.5 | 20 | 125 |

Close fitting butt welds were produced with welding conditions listed in Table 3. After welding, the specimens were firstly inspected using X-ray radiography to detect any porosity in the welds. Secondly, metallographic samples were cut from the specimen, then ground, polished and finally etched in a solution of 2 mL of HF + 10 mL of HNO$_3$ + 88 mL of water. The shapes and dimensions of the welds were observed and the weld profile defects were measured with an optical microscope (OM) (Olympus, Tokyo, Japan). For clarity, Figure 3 details the definition of the parameters and defects relating to weld profiles. Thirdly, static tensile tests were carried out to evaluate the mechanical properties of the welds, and fracture surfaces were examined with scanning electron microscope (SEM) (FEI Company, Hillsboro, OR, USA) to determine the fracture features. Finally, the influences of the two welding positions on the weld morphologies, porosity levels, and mechanical properties were analyzed.

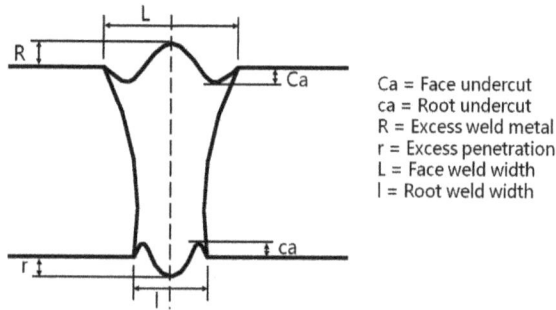

Ca = Face undercut
ca = Root undercut
R = Excess weld metal
r = Excess penetration
L = Face weld width
l = Root weld width

**Figure 3.** Weld profile parameters and defects.

## 3. Results

### 3.1. Weld Profile Defects

The cross sections of the laser welds produced with the two sets of laser welding parameters used are shown in Figures 4 and 5, respectively. The dimensions of weld profile defects measured for these laser welds are listed in Tables 4 and 5.

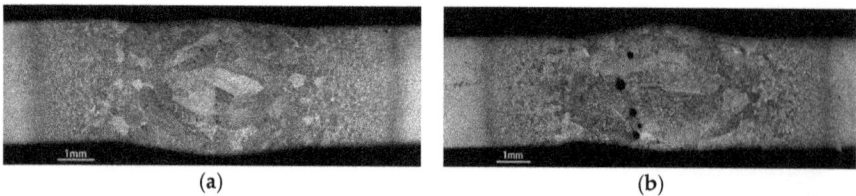

(a)

(b)

**Figure 4.** Cross sections of laser welds (laser power: 2.2 kW; welding speed: 8 mm/s). (**a**) Flat welding; (**b**) Horizontal welding.

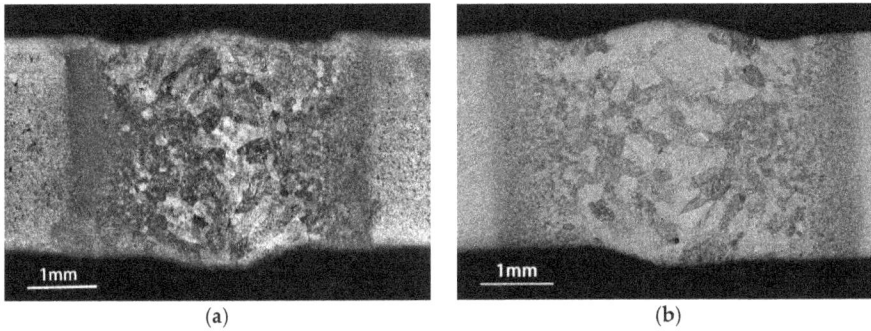

Figure 5. Cross sections of laser welds (laser power: 2.5 kW; traveling speed: 20 mm/s). (a) Flat welding; (b) Horizontal welding.

Table 4. Undercuts of welds produced under different welding conditions (unit: μm).

| Welding Position | Butt Weld Identity | Face Undercut | | Root Undercut | | Sum of Face and Root Undercuts | |
|---|---|---|---|---|---|---|---|
| | | Left Side | Right Side | Left Side | Right Side | Left Side | Right Side |
| Flat | F01 | 71.1 | 75.8 | 0 | 0 | 71.1 | 75.8 |
| | F02 | 116.1 | 101.9 | 0 | 23.7 | 116.1 | 125.6 |
| Horizontal | H01 | 81.0 | 0 | 111.9 | 25.0 | 192.9 | 25 |
| | H02 | 116.0 | 32.9 | 46.5 | 0 | 162.5 | 32.9 |

Table 5. Excess weld metals, excess penetrations, face and root weld widths for different welding conditions (unit: μm).

| Welding Position | Butt Weld Identity | Excess Weld Metal | Excess Penetration | Face Weld Width | Root Weld Width |
|---|---|---|---|---|---|
| Flat | F01 | 11.8 | 252.4 | 6194.3 | 6111.4 |
| | F02 | 86.1 | 314.0 | 3784.4 | 3265.4 |
| Horizontal | H01 | 255.8 | 100.3 | 6546.6 | 6339.7 |
| | H02 | 199.1 | 173.8 | 4169.0 | 3470.7 |

Table 4 lists the undercuts of welds, from which it can be found that, for flat welds, the face undercut is greater than the root undercut. Undercuts on the left side and the right side are essentially the same, i.e., symmetrical about the weld centerline. A decrease in heat input (i.e., using 2.5 kW and 20 mm/s, instead of 2.2 kW and 8 mm/s) results in a noticeable increase in face undercut, while the root undercut is less affected. For horizontal welds, the undercuts on the left side and the right side are no longer symmetrical. The left side (which is on the top side during horizontal welding) is notably larger than the right side (which is on the bottom side during welding). A decrease in the heat input results in an increase in the face undercut but decrease in the root undercut. On the whole, flat welding leads to larger face undercuts but smaller root undercuts than horizontal welding. With regard to the sum of the face and root undercuts on the same side (i.e., left side or right side), it can be found that the sum on the left side for the horizontal welds is notably larger than that on the right side, and is also larger than both the left and right side for the flat welds.

The excess weld metal and excess penetration values are presented in Table 5. Among all welds produced in this study, the flat welds have the least excess weld metals but noticeable excess penetrations. Large differences exist between the excess weld metal and the excess penetration for flat welds. The horizontal welds have larger excess metal values than excess penetration values. The differences between the excess weld metal values and the excess penetrations are smaller than the differences for the flat welds.

The face and root weld widths of welds are also given in Table 5. For both sets of welding parameters, the face and root widths for the flat welds are slightly smaller than those for the horizontal welds. Face weld widths are larger than root weld widths for both flat welds and horizontal welds. A decrease in the heat input results in decreases in both the face weld width and root weld width, as would be anticipated.

*3.2. Porosity in Welds*

Figure 6 shows the X-ray radiographs of four laser butt welds produced with the two different welding positions and the two different sets of welding parameters.

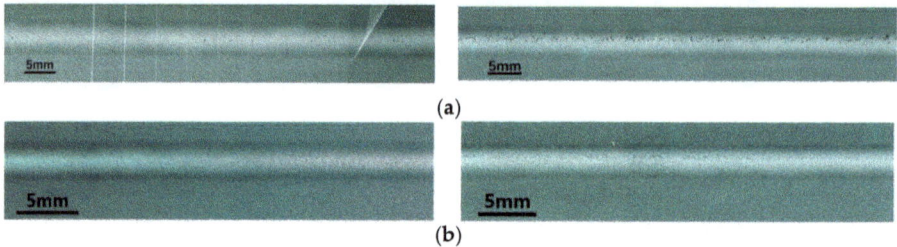

**Figure 6.** X-ray radiographs of butt welds made under flat and horizontal welding positions with two different sets of welding parameters. (**a**) Flat welding (**left**) versus horizontal welding (**right**) for 2.2 kW and 8 mm/s; (**b**) Flat welding (**left**) versus horizontal welding (**right**) for 2.5 kW and 20 mm/s.

For welding parameters with higher heat input (Figure 6a, heat input 275 J/mm), several isolated pores can be found along the centerline of the flat weld. By contrast, extensive chain porosity is detected above the centerline of the corresponding horizontal weld.

For welding parameters with lower heat input (Figure 6b, 125 J/mm), a limited amount of porosity can be seen in the flat weld. In the horizontal weld, fine scale chain porosity is detected. The pores are very close to each other and become indistinguishable individually, and they distribute in the upper half part of the weld.

For a quantitative characterization, the cumulative lengths of porosity over a 100 mm weld length were measured and are presented in Figure 7. It is confirmed that the porosity contents for the flat weld are lower than those in the horizontal welds. For both flat and horizontal welding positions, using higher laser power and correspondingly higher welding speed (2.5 kW, 20 mm/s) can also help to reduce the porosity content.

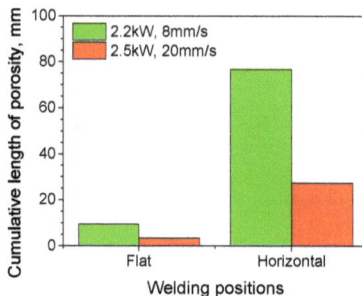

**Figure 7.** Cumulative lengths of porosity in welds with different welding conditions.

## 3.3. Mechanical Behavior

Static tensile tests showed that, irrespective of the welding position used, specimens fractured through the weld metals in welds with the lower heat input (2.5 kW and 20 mm/s), and through the base metals in welds made with higher heat input (2.2 kW and 8 mm/s).

Figure 8 shows the static tensile properties of the laser welds made using four welding conditions. It can be seen that, when a lower heat input (2.5 kW, 20 mm/s) was employed, comparable tensile strengths and specific elongations were achieved for flat welds and horizontal welds: these welds fractured through the base metals owing to the higher strengths of the weld metals compared to that of the base metal. The yield strengths of the flat welds made with the higher heat input (2.2 kW, 8 mm/s) were similar to those made with the lower heat input. By contrast, the yield strengths of the horizontal welds decreased by about 25 MPa when the higher heat input conditions were used.

In addition, marked decreases in specific elongation could be seen in both flat and horizontal welds when the higher energy input was used. Such deterioration in ductility was more significant for horizontal welds than for flat welds in these cases.

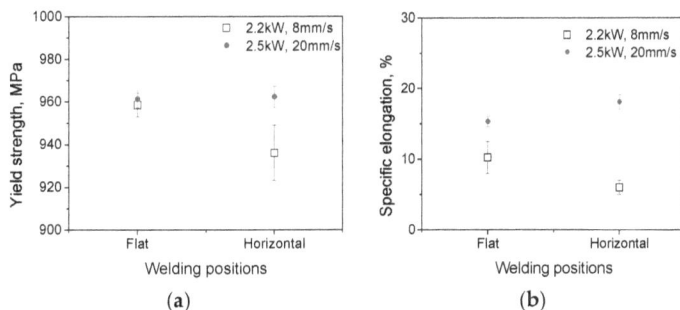

**Figure 8.** Static tensile properties of the laser welds made using four welding conditions. (**a**) Yield strength; (**b**) Specific elongation.

# 4. Discussion

## 4.1. Effects of Gravity on Weld Profile

Results on weld profile defects show that using the flat welding position leads to deeper undercuts on the top side than the bottom side of the weld, while using horizontal welding position leads to deeper undercuts on the left side (on the top side during welding) than on the right side (on the bottom side during welding) of the weld.

These differences can be attributed to enhanced fluid flow towards the root of the weld under the action of gravity for the flat welding position (as shown in Figure 9a); for the horizontal welding position, gravity drives the molten metal flow from the top side of the weld toward the bottom side, leading to necking on the top and thickening on the bottom (as shown in Figure 9b). Such a lateral flow also contributes to horizontal welds being wider than flat welds for a given set of welding parameters. The notable downward movement of the weld pool metal when the flat position welding also accounts for the smaller amount of excess weld metal and larger amount of excess penetration, when compared with horizontal welds.

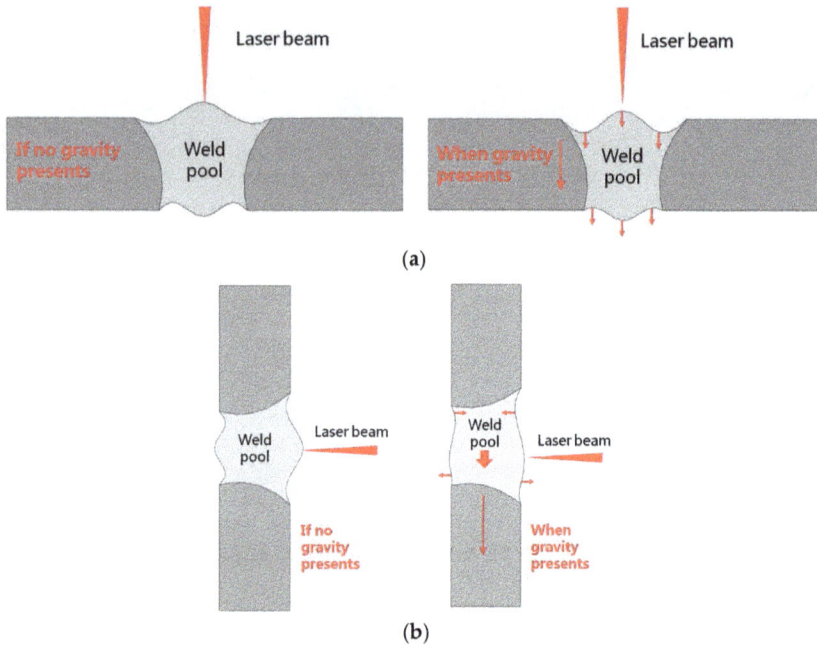

**Figure 9.** Schematic representation of the molten metal flow and movement of the weld pool surface due to the action of gravity in (**a**) the flat position and (**b**) the horizontal position.

*4.2. Effects of Gravity on Porosity*

Porosity content results indicated that using the flat welding position will result in less porosity when compared with the horizontal welding position. This can be attributed to the different degree of ease with which the pores can escape from the solidifying weld pool under different welding positions. As shown in Figure 10a, in flat welding, the pores formed in the weld pool will float upwards under the action of melt flow and buoyancy, which has been observed using an X-ray transmission imaging system by Katayama et al. [13,14]. Some pores can escape through the top surface of the weld pool before it solidifies, while others that are not able to escape will remain in the welds and form porosity. For horizontal welding, shown in Figure 10b, the pores formed also float upwards and move away from the weld centerline. However, the uppermost surface of the weld pool is now in contact with the unmelted base rather than free space. This restricts those pores from escaping; consequently, almost all end up entrapped within the weld bead. Therefore, a high porosity content, located above the weld centerline, exists within the horizontal welds, as revealed in Figure 6, which is also visible from the cross section of the horizontal weld shown in Figure 4b.

Furthermore, the final amount of porosity in a weld, especially when horizontal welding, will depend on the amount of pores formed during welding, which is closely related to the stability of the keyhole and the fluid flow characteristics in the weld pool. Results (Figure 6) show that higher power (2.5 kW) at higher welding speed (20 mm/s) result in less porosity compared with lower power (2.2 kW) at lower speed (8 mm/s). This agrees with results reported previously [15], which indicated through computational fluid dynamic (CFD) modeling that the fluid flow behind the keyhole can be quite unstable, and vortices and pores can form, particularly more so when using lower power and lower welding speed conditions. In contrast, when using higher power and welding speed conditions, CFD modelling predicted that the fluid flow would be less unstable, and fewer pores would result.

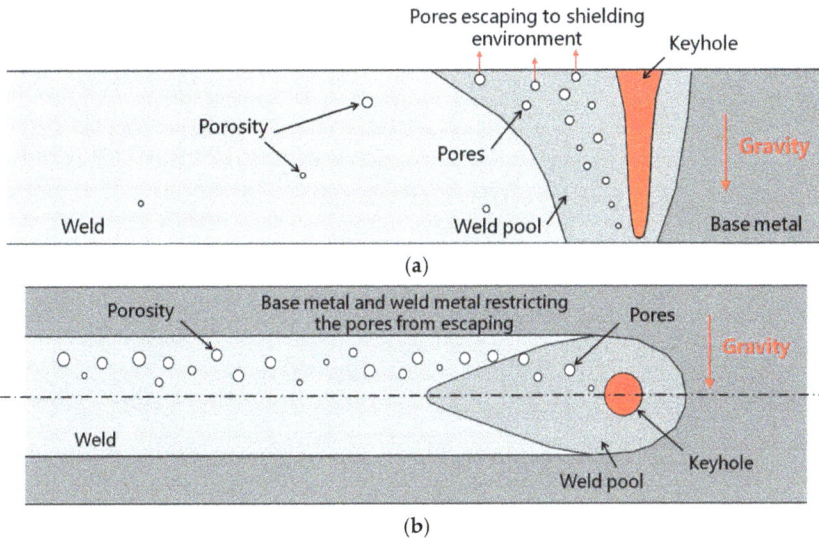

**Figure 10.** Schematic diagrams showing the movement of pores for (**a**) the flat position and (**b**) the horizontal position welding.

*4.3. Effects of Welding Position on Static Tensile Properties*

For laser welds made in Ti6Al4V, weld toe undercuts may result in high stress concentrations, which in turn lead to crack initiation and propagation during tensile testing [16]. This seems true in this work in the case of the horizontal welds made with higher heat input (2.2 kW, 8 mm/s), which had the largest undercut. However, this was not the case for the flat welds made with the same heat input, which still also fractured through the weld metal, in spite of their smaller undercuts (ref. Table 4). Evidently, the fracture position is determined not only by the degree of undercut but also other factors. It has been indicated that higher heat input will make the microstructure in the (martensitic) weld coarser, which then gives a lower toughness and a greater tendency toward cracking [16]. This could explain why all of the welds produced with a higher heat input failed through the weld metal, while those made with lower heat input failed through the base metal.

For welding conditions with higher heat input (2.2 kW, 8 mm/s), a greater amount of porosity is formed (ref. Figure 6), and this may also deteriorate weld strength. This strength deterioration is not noted in flat welds, because their porosity levels are low overall. By contrast, a great number of pores are entrapped in horizontal welds, which could then explain the notable decrease in weld strength. Figure 11 shows the comparative fractographs of a horizontal and a flat weld, both made with higher heat inputs. As Figure 11 shows, large pores are seen in the fracture face of the horizontal weld, while porosity is not detected in that of the flat weld.

Irrespective of welding position, all welds that fractured through the base metal had greater ductility than those that fractured through the weld metal. This is because the martensite of the weld metal has worse ductility than the phases ($\alpha + \beta$) of the base metal Ti6Al4V, as previously reported [17–23]. In addition, as porosity is to the detriment of the ductility of welds when its amount reaches a certain limit [24], the great amount of porosity present in the horizontal welds could make the ductility even worse.

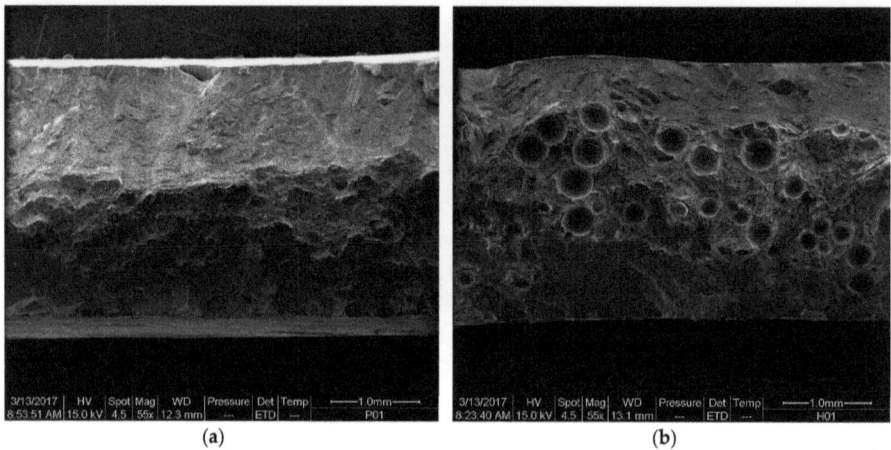

**Figure 11.** Fractographs of the horizontal weld (**a**) and the flat weld (**b**) made with higher heat input (2.2 kW, 8 mm/s).

## 5. Conclusions

The following conclusions have been drawn from work carried out:

(1) For flat welds, face undercut was larger than root undercut; for horizontal welds, the undercut on the left side (top side during welding) side was larger than that on the right side (bottom side during welding).

(2) The excess penetration was greater than the excess weld metal for the flat welds, while for the horizontal welds, the excess penetration was smaller than the excess weld metal. For the same laser welding parameters, the horizontal welds were wider than the flat welds.

(3) The horizontal welding position resulted in higher weld metal porosity contents than the flat welding position, because there were fewer routes for pores to escape from the weld pool during horizontal welding. The pores were located in the center plane of the flat welds, and above the center plane of the horizontal welds.

(4) In the welds investigated, the undercuts did not show an association with the fracture positions nor the strengths in static tensile testing, although excessive porosity in laser welds did lead to significant decreases in their strength and specific elongation.

(5) Compared with a horizontal welding position, the flat welding position led to better weld formation, less porosity, and higher tensile strength. For both flat and horizontal welding positions, it is recommended to use higher laser powers and welding speeds to reduce weld porosity and improve the mechanical properties of laser welds in Ti6Al4V alloys.

**Acknowledgments:** This work has been financially supported by the National Natural Science Foundation of China (www.nsfc.gov.cn, No. U1537205 and No. 51675303) and the Tsinghua University Initiative Scientific Research Program (No. 2014Z05093).

**Author Contributions:** B.C. and D.D. conceived and designed the experiments; H.P. and H.C. performed the experiments; Z.Y. and J.S. analyzed the data; H.L. contributed reagents/materials/analysis tools; B.C. wrote the paper.

**Conflicts of Interest:** The authors declare no conflict of interest.

## References

1. Leyens, C.; Peters, M. *Titanium and Titanium Alloys*; WILEY-VCH Verlag GmbH & Co. KGaA: Weinheim, Germany, 2003.

2. Chang, B.; Blackburn, J.; Allen, C.; Hilton, P. Studies on the spatter behaviour when welding AA5083 with a Yb-fibre laser. *Int. J. Adv. Manuf. Technol.* **2016**, *84*, 1769–1776. [CrossRef]
3. Guo, W.; Liu, Q.; Francis, J.A.; Crowther, D.; Thompson, A.; Liu, Z. Comparison of laser welds in thick section S700 high-strength steel manufactured in flat (1G) and horizontal (2G) positions. *CIRP Ann. Manuf. Technol.* **2015**, *64*, 197–200. [CrossRef]
4. Shen, X.; Li, L.; Guo, W.; Teng, W.; He, W. Comparison of processing window and porosity distribution in laser welding of 10 mm thick 30CrMnSiA ultrahigh strength between flat (1G) and horizontal (2G) positions. *J. Laser Appl.* **2016**, *28*. [CrossRef]
5. Sohail, M.; Han, S.-W.; Na, S.-J.; Gumenyuk, A.; Rethmeier, M. Numerical investigation of energy input characteristics for high-power fiber laser welding at different positions. *Int. J. Adv. Manuf. Technol.* **2015**, *80*, 931–946. [CrossRef]
6. Kumar, A.; Debroy, T. Heat transfer and fluid flow during gas-metal-arc fillet welding for various joint configurations and welding positions. *Metall. Mater. Trans. A* **2007**, *38*, 506–519. [CrossRef]
7. Cho, D.W.; Na, S.J.; Cho, M.H.; Lee, J.S. A study on V-groove GMAW for various welding positions. *J. Mater. Process. Technol.* **2013**, *213*, 1640–1652. [CrossRef]
8. Cai, X.Y.; Fan, C.L.; Lin, S.B.; Yang, C.L.; Bai, J.Y. Molten pool behaviors and weld forming characteristics of all-position tandem narrow gap GMAW. *Int. J. Adv. Manuf. Technol.* **2016**, *87*, 2437–2444. [CrossRef]
9. Xu, W.H.; Lin, S.B.; Fan, C.L.; Yang, C.L. Prediction and optimization of weld bead geometry in oscillating arc narrow gap all-position GMA welding. *Int. J. Adv. Manuf. Technol.* **2015**, *79*, 183–196. [CrossRef]
10. Xu, W.H.; Lin, S.B.; Fan, C.L.; Zhuo, X.Q.; Yang, C.L. Statistical modelling of weld bead geometry in oscillating arc narrow gap all-position GMA welding. *Int. J. Adv. Manuf. Technol.* **2014**, *72*, 1705–1716. [CrossRef]
11. Chen, Y.B.; Feng, J.C.; Li, L.Q.; Li, Y.; Chang, S. Effects of welding positions on droplet transfer in $CO_2$ laser-MAG hybrid welding. *Int. J. Adv. Manuf. Technol.* **2013**, *68*, 1351–1359. [CrossRef]
12. Koga, S.; Inuzuka, M.; Nagatani, H.; Iwase, T.; Masuda, H.; Ushio, M. Study of all position electron beam welding process for pipeline joints. *Sci. Technol. Weld. Join.* **2000**, *5*, 105–112. [CrossRef]
13. Katayama, S.; Mizutani, M.; Matsunawa, A. Development of porosity prevention procedures during laser welding. *Proc. SPIE* **2003**, *4831*, 281–288.
14. Katayama, S.; Seto, N.; Mizutani, M.; Matsunawa, A. Formation mechanism of porosity in high power YAG laser welding. In *Laser Materials Processing, ICALEO 2000 Proceedings*; Taylor & Francis: Abingdon, UK, 2000; Volume 89, pp. 16–25.
15. Chang, B.; Allen, C.; Blackburn, J.; Hilton, P.; Du, D. Fluid flow characteristics and porosity behavior in full penetration laser welding of a titanium alloy. *Metall. Mater. Trans B* **2015**, *46*, 906–918. [CrossRef]
16. Yang, J.; Cheng, D.; Huang, J.; Zhang, H.; Zhao, X.; Guo, H. Microstructure and mechanical properties of Ti-6AL-4V joints by laser beam welding. *Rare Metal Mater. Eng.* **2009**, *38*, 259–262. (In Chinese).
17. Mazumder, J.; Steen, W.M. Microstructure and mechanical properties of laser welded titanium 6Al-4V. *Metall. Trans. A* **1982**, *13A*, 865–871. [CrossRef]
18. Squillace, A.; Prisco, U.; Ciliberto, S.; Astarita, A. Effect of welding parameters on morphology and mechanical properties of Ti-6Al-4V laser beam welded butt joints. *J. Mater. Process. Technol.* **2012**, *212*, 427–436. [CrossRef]
19. Fang, X.; Zhang, J. Microstructural evolution and mechanical properties in laser beam welds of Ti–2Al–1.5Mn titanium alloy with transversal pre-extrusion load. *Int. J. Adv. Manuf. Technol.* **2016**, *85*, 337–343. [CrossRef]
20. Junaida, M.; Baigb, M.N.; Shamirc, M.; Khand, F.N.; Rehmana, K.; Haiderea, J. A comparative study of pulsed laser and pulsed TIG welding ofTi-5Al-2.5Sn titanium alloy sheet. *J. Mater. Process. Technol.* **2017**, *242*, 24–38. [CrossRef]
21. Hong, K.-M.; Shin, Y. C. Analysis of microstructure and mechanical properties change in laser welding of Ti6Al4V with a multiphysics prediction model. *J. Mater. Process. Technol.* **2016**, *237*, 420–429. [CrossRef]
22. Kashaev, N.; Ventzke, V.; Fomichev, V.; Fomin, F.; Riekehr, S. Effect of Nd:YAG laser beam welding on weld morphology and mechanical properties of Ti-6Al-4V butt joints and T-joints. *Opt. Lasers Eng.* **2016**, *86*, 172–180. [CrossRef]

23. Zhang, K.; Ni, L.; Lei, Z.; Chen, Y.; Hu, X. Microstructure and tensile properties of laser welded dissimilar Ti-22Al-27Nb and TA15 joints. *Int. J. Adv. Manuf. Technol.* **2016**, *87*, 1685–1692. [CrossRef]
24. Zhang, Y.; Huo, L.; Jing, H.; Pan, R. The effects of porosities and slag inclusions on mechanical properties of welded joint. *Press. Vessels* **1996**, *13*, 34–38. (In Chinese).

![applied sciences logo](applied sciences)

MDPI

*Article*

# Effect of Molybdenum on the Microstructures and Properties of Stainless Steel Coatings by Laser Cladding

Kaiming Wang [1], Baohua Chang [1,*], Jiongshen Chen [2], Hanguang Fu [2,*], Yinghua Lin [3] and Yongping Lei [2]

[1]   State Key Laboratory of Tribology, Department of Mechanical Engineering, Tsinghua University, Haidian District, Beijing 100084, China; kmwangbjut@163.com
[2]   School of Materials Science and Engineering, Beijing University of Technology, Number 100, Pingle Garden, Chaoyang District, Beijing 100124, China; S201709128@emails.bjut.edu.cn (J.C.); yplei@bjut.edu.cn (Y.L.)
[3]   Institute of Laser Advanced Manufacturing, Zhejiang University of Technology, Hangzhou 310014, Zhejiang, China; lyh351258@163.com
*   Correspondence: bhchang@tsinghua.edu.cn (B.C.); hgfu@bjut.edu.cn (H.F.)

Received: 11 September 2017; Accepted: 11 October 2017; Published: 15 October 2017

**Abstract:** Stainless steel powders with different molybdenum (Mo) contents were deposited on the substrate surface of 45 steel using a 6 kW fiber laser. The microstructure, phase, microhardness, wear properties, and corrosion resistance of coatings with different Mo contents were studied by scanning electron microscopy (SEM), electron probe microanalyzer (EPMA), X-ray diffraction (XRD), microhardness tester, wear tester, and electrochemical techniques. The results show that good metallurgical bonding was achieved between the stainless steel coating and the substrate. The amount of $M_7(C, B)_3$ type borocarbide decreases and that of $M_2B$ and $M_{23}(C, B)_6$ type borocarbides increases with the increase of Mo content in the coatings. The amount of martensite decreases, while the amount of ferrite gradually increases with the increase of Mo content. When the Mo content is M, $Mo_2C$ phase appears in the coating. The microstructure of the coating containing Mo is finer than that of the Mo-free coating. The microhardness decreases and the wear resistance of the coating gradually improves with the increase of Mo content. The wear resistance of the 6.0 wt. % Mo coating is about 3.7 times that of the Mo-free coating. With the increase of Mo content, the corrosion resistance of the coating firstly increases and then decreases. When the Mo content is 2.0 wt. %, the coating has the best corrosion resistance.

**Keywords:** laser cladding; stainless steel powder; Mo content; microstructure; wear resistance; corrosion resistance

---

## 1. Introduction

Carbon steel and low alloy steel, which have a high strength, good processing performance, and low price, are widely used in marine environments [1,2]. Due to the complex environment of the ocean, some damage will inevitably occur on the surfaces of steels. The main failure ways of steel surfaces include wear, corrosion, and oxidation [3–5]. Hence, surface modification technologies are necessary to improve the performance of steel parts. The coatings cladded on the surface of steels can greatly improve the service life of steel facilities in the ocean, which have a significant influence on the reliability of offshore steel components and can enhance the economic efficiency of the ocean [6].

Laser cladding technology is a new kind of surface modification technology. Coatings with specific properties can be obtained by melting alloy powder on the surface of a substrate through a laser beam [7]. Compared to the traditional surface modification technologies such as surfacing welding, thermal spraying, etc., the laser cladding process can control the heat input more precisely, has basically no limits on the alloy powders that can be used, and is suitable for flexible processing. The coatings by laser cladding have a dense microstructure, lower dilution rate, and less thermal deformation [8–11]. In addition, laser cladding has high material utilization and low energy consumption, and it is considered as an environmentally friendly materials processing technology. Therefore, the laser cladding has been widely studied and used for surface modification in recent years [12]. Tan et al. [13] researched the effect of different $Al/Fe_2O_3$ thermite reactants on the Fe-based coating in laser cladding. The results showed that $Al_2O_3$ ceramic and $M_7C_3$ type carbide were in situ synthesized by laser cladding. With the increase of thermite reactants, the amount of $Al_2O_3$ ceramic and $M_7C_3$ carbide in the coatings increased gradually. The microhardness and the wear resistance of the coatings could be improved when increasing the amount of thermite reactant. Ray et al. [14] used three different Ni-based alloy powders to strengthen the surface of a continuous casting roller by laser cladding. The on-line results showed that the continuous casting roller exhibited a better performance after being reinforced by the laser cladding.

Powders used in laser cladding mainly include Fe-based, Ni-based, and Co-based alloy powders [15–17]. The Fe-based alloy is widely available and inexpensive, and it is the most widely used material in industrial production [18]. A good wetting performance and high bonding strength can be achieved between the coating and substrate by using Fe-based alloy powders on the surface of steel substrates in laser cladding [19–21]. Wang et al. [22] prepared an FeCrBSi coating on the surface of 45 steel by laser cladding. The toughness of the coating was improved and the crack tendency was reduced by increasing the Cr content. The average microhardnesses varied from 760 to 950 HV. Adding a specific alloy component in the alloy powder could further improve the properties of the laser cladding layer [23]. Incorporating hard particles like WC [24], VC [25], TiC [26], and TiB [27] in the coating could improve the wear resistance of the coating. Rare earth elements like $CeO_2$ [28] and $Y_2O_3$ [29] could refine the grain and improve the properties of the coating. Mo could refine the grain, increase the toughness, enhance the plasticity, and reduce the crack sensitivity of the coating, and was beneficial for improving the performance of the coating [30,31]. Ding et al. [32] studied the effect of Mo on the microstructure and wear resistance of laser cladded Ni-based alloys on Q235 steel. The results indicated that Mo could refine grains and synthesize polygonal equiaxed grains. The microhardness and wear resistance of Ni-based coatings were improved greatly by the addition of a moderate amount of Mo.

Although the beneficial effect of Mo on the coating has already been demonstrated for some alloy powders, the influence of Mo content on the performance of the laser cladded coating of the stainless steel alloy powder has not been well investigated up to date. In this paper, the stainless steel coatings with different Mo contents were laser cladded on the surface of 45 steel, and the microstructure, hardness, wear resistance, and corrosion resistance of the coatings were studied. The obtained results could provide theoretical guidance for the industrial applications of stainless steel coatings.

## 2. Materials and Methods

### 2.1. Specimen Preparation

Laser cladding was carried out using a fiber laser cladding system, which consisted of an IPG YLS-6000 fiber laser (IPG Photonics Corporation, Oxford, MA, USA), powder feeding system, cooling system, and control system. The laser head was integrated with an ABB-4600(Asea Brown Boveri)robot (Asea Brown Boveri Ltd., Zurich, Switzerland). A DPSF-2 powder feeding system (Beijing Institute of Aeronautical Manufacturing, Beijing China) and a coaxial nozzle (HIGHYAG Corporation, Berlin, Germany) were used to feed powders into the molten pool by argon gas. The gas flow was 15 L/min and the purity of argon gas was $\geq$ 99.9%. The substrate material was 45 steel (Baosteel Corporation, Shanghai, China). The nominal composition of the 45 steel plate is shown in Table 1. The dimension of the specimen was 150 mm $\times$ 150 mm $\times$ 12 mm. The surface of the substrate was abraded on SiC grit paper (Suzhou suboli grinding material Co. Ltd, Suzhou, China) with the mesh of 200 # and 400 #, and acetone was then used to clean the surface of the substrate before the experiment. The cladding alloy material was 431 stainless steel alloy powder (Höganäs Corporation, Skåne, Sweden) with a particle size of 50–105 µm, and the nominal composition of 431 stainless steel alloy powder is shown in Table 1. The Molybdenum element for the experiment was provided by Mo-Fe powder (Qinghe Xinbao Alloy Material Co., Ltd, Xingtai, China). The composition of Mo-Fe powder is as follows: 50.30 wt. % Mo, 0.37 wt. % C, 0.031 wt. % P, 0.016 wt. % S, Fe: Bal. The size of the Mo-Fe powder was less than 50 µm. In this paper, the Mo-Fe powder with the mass fraction of 4%, 8% and 12% was mixed with 431 alloy powder, and the Mo content in the composite powder was 2%, 4%, and 6%, respectively. In order to ensure that the powder was mixed evenly, the composite powders were mixed using the ball mill (Changsha planetary machinery and equipment factory, Changsha, China). The milling speed was 160 r/min and milling time was 11 h. The size and morphology of the 431 and Mo-Fe powders were basically not changed by the ball milling process. The powders were placed in a drying stove (Suzhou Jiangdong precision instruments Co., Ltd., Suzhou, China) and dried at 100 °C for about 2 h before the laser cladding experiments.

**Table 1.** Chemical composition of the 45 steel substrate and 431 stainless steel alloy powder (wt. %).

| Element | C | Cr | Ni | Si | Mn | B | Co | Fe |
|---|---|---|---|---|---|---|---|---|
| 45 steel | 0.44 | 0.15 | 0.15 | 0.24 | 0.69 | - | - | Bal. |
| 431 powder | 0.19 | 18.44 | 2.40 | 0.75 | 0.26 | 0.87 | 0.53 | Bal. |

On the basis of previous experiments [33,34], the laser cladding process parameters were chosen as follows: the laser power, the scanning speed, and the feeding powder rate were 2000 W, 240 mm/min, and 15 g/min, respectively. Specimens for metallographic examinations were cut, ground, and polished according to standard procedures and etched with a solution consisting of HCl (3 mL) (Beijing Chemical Works, Beijing, China) and $HNO_3$ (1 mL) (Beijing Chemical Works, Beijing, China) for 50 s. After the completion of etching, the residual etching solution was washed with water, and the surface was then dried with a blower (PHILIPS, Amsterdam, Netherlands).

### 2.2. Microstructure Observation

The microstructure of the coating was investigated by an OLYMPUS BX51 optical metallographic microscope (OM) (OLYMPUS Corporation, Tokyo, Japan) and Quanta FEG650 scanning electron microscope (SEM) (FEI Company, Hillsboro, OR, USA). Element distributions were analyzed by energy disperse spectroscopy (EDS) (FEI Company, Hillsboro, OR, USA) and an electron probe microanalyzer (EPMA) (SHIMADZU Corporation, Kyoto, Japan). The phase identification of the coatings was carried out on a Shimadzu XRD-7000 X-ray diffractometer (XRD) (SHIMADZU Corporation, Kyoto, Japan). The detailed parameters of XRD were as follows: Cu-K$_\alpha$ radiation at 40 kV and 200 mA as an X-ray source. Specimens were scanned in the 2θ range of 20°~80° in a step-scan mode (0.02° per step).

### 2.3. Hardness and Wear Resistance Test

The microhardness of the coating was measured by a MICRO MET-5103 digital microhardness tester (Shanghai Nazhi Electronic Technology Co., Ltd, Shanghai, China) with a load of 5 N and holding time for 10 s. Each sample was measured and the microhardness of the coating was the average of ten measurements.

The wear test was carried out on an M-200 wear test machine, of which the schematic diagram is shown in Figure 1. The size of the sample was 10 mm × 10 mm × 12 mm and the grinding ring material was GCr15 steel (its chemical composition: 1.03% C, 1.49% Cr, 0.35% Mn, 0.27% Si). The hardness of GCr15 steel was 60.5 HRC. The detailed measuring parameters of the wear test were as follows: the test load 294 N, the test machine speed 200 rpm, the experiment temperature 20 °C, the test time 30 min. The weight loss of the sample was weighed by TG328B balance (Shanghai Liangping Co., Ltd , Shanghai, China). The weighing range and the precision of the balance were 200 g and 0.1 mg, respectively the wear resistance of the coating was the ratio of wear time/weight loss. The greater the value, the better the wear resistance of the coating. The microstructure of the worn surface was investigated by Quanta FEG650 SEM.

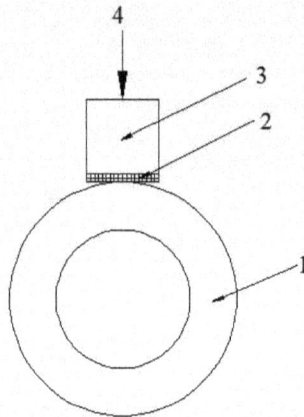

**Figure 1.** Schematic diagram of wear tester. 1—Rotating ring, 2—The coating, 3—45 steel substrate, 4—Load.

*2.4. Electrochemical Corrosion*

The surfaces of the coating were machined to a size of 12 mm × 12 mm using a wire cutting machine, and were then well polished. Electrochemical measurements in 3.5% NaCl (Beijing Chemical Works, Beijing, China) solutions were carried out on an electrochemical workstation (Parstat 2273, AMETEK, Berwyn, PA, USA) at 25 °C. The potentiodynamic polarization curves were measured at a scanning speed of 0.5 mV/s from −1.0 to 0.5 V. The conventional three-electrode cell was used, with the coating or matrix sample as the working electrode, the Ag/AgCl electrode as the reference electrode, and the platinum electrode as the counter electrode.

## 3. Results and Discussion

*3.1. Microstructure Analysis*

In order to identify the phase composition of the coatings with different Mo contents, XRD analysis were conducted on the coating. The XRD spectra are presented in Figure 2. With the increase of Mo content, the phases in the coating are changed and the intensity of the diffraction peak becomes weaker. The phases in the Mo-free coating are $\alpha$-Fe, $M_2B$ (M = Fe, Ni), $M_{23}(C, B)_6$, and $M_7(C, B)_3$. Another $\alpha$-Fe peak appears near the main peak when Mo is added into the coating. The intensity of the peak gradually increases with the increase of Mo content. When the Mo content is 4.0 wt. %, $Mo_2C$ phase appears in the coating. With the increase of Mo content, the amount of $M_7(C, B)_3$ decreases gradually, and the amount of $M_2B$, $M_{23}(C, B)_6$, and $Mo_2C$ gradually increases.

Figure 3 shows the metallographic microstructure of the coating. It can be seen from the bottom part of the coating (Figure 3a,c) that a light line is formed between the coating and the substrate. The light line is the fusion that grows mainly in the form of planar grains. The fusion line indicates a good metallurgical bonding between the substrate and the coating. During the solidification process, the microstructure of the coating is mainly determined by the ratio (G/R) of the temperature gradient (G) and solidification rate (R) [35]. The temperature gradient (G) is high and the solidification rate (R) is very low at the bottom of the molten pool, so the ratio of G/R is relatively large and the solidification structure grows as planar grains [36]. With the increase of distance from the fusion line, the temperature gradient (G) decreases and the solidification rate (R) increases, so that the G/R value is reduced, and the microstructure of the coating is transformed from planar grains to cell grains, dendrites, and equiaxed grains [37].

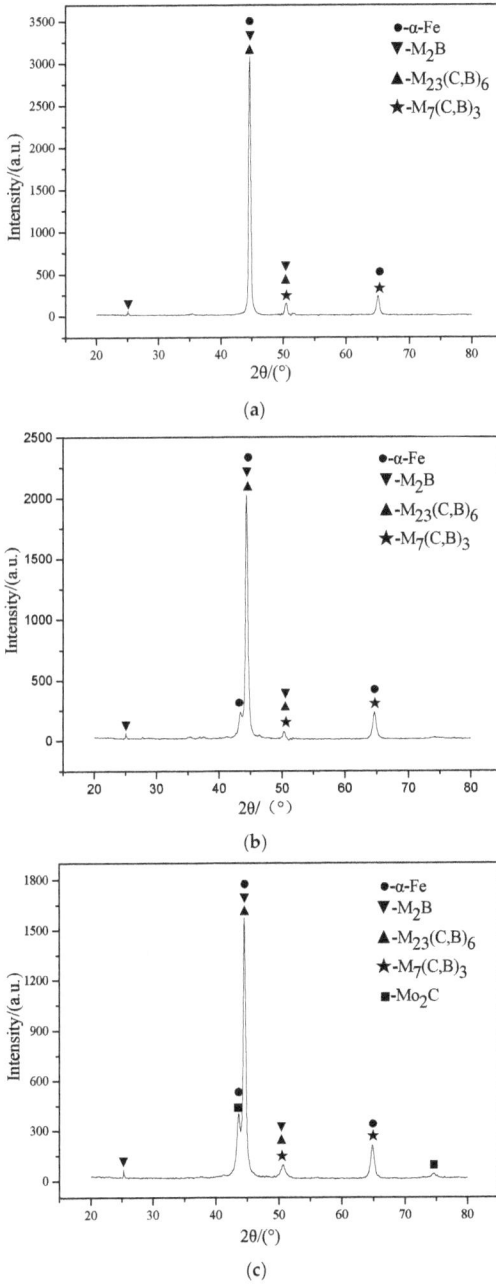

(a)

(b)

(c)

**Figure 2.** *Cont.*

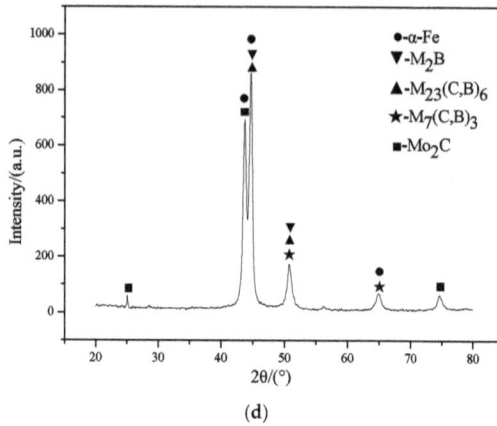

(d)

**Figure 2.** XRD spectrums of coatings: (**a**) Mo-free; (**b**) 2.0 wt. % Mo; (**c**) 4.0 wt. % Mo; (**d**) 6.0 wt. % Mo.

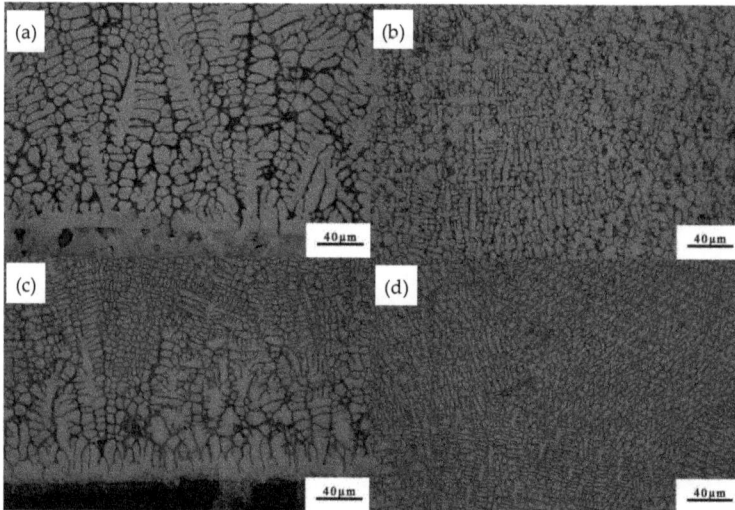

**Figure 3.** Optical micrographs of the coating: (**a**) The bottom part of the Mo-free coating; (**b**) The upper part of the Mo-free coating; (**c**) The bottom part of the 4.0 wt. % Mo coating; (**d**) The upper part of the 4.0 wt. % Mo coating.

Figure 4 shows the scanning electron micrographs of the coating with different Mo contents. It can be seen from Figure 4 that the coating is mainly composed of a netlike eutectic structure, small granular eutectic structure, and dendrite matrix. During the solidification process, the dendrite matrix is firstly precipitated from the liquid phase, and then the eutectic structure between the dendrites is formed through the eutectic reaction in the remaining liquid phase. The size of the dendrite is obviously refined with the increase of Mo content from 0 to 6.0 wt. %, as shown in Figure 4. The sizes of the dendrite are measured by Image Pro-plus 6.0 software, and the average sizes of dendrites in the coatings of Mo-free, 2.0 wt. % Mo, 4.0 wt. % Mo, and 6.0 wt. % Mo are 7.9 μm, 6.5 μm, 4.4 μm, and 3.9 μm, respectively. This may result from the addition of Mo, which affects the nucleation process of the coating [38]. On the one hand, the Mo element can prevent the growth of austenite, which can refine the structure of the coating. On the other hand, part of Mo can dissolve in the coating, which results in lattice distortion. The compositions of netlike eutectic structure 1, the dendrite structure 2, and the granular eutectic structure 3 with the addition of 4.0 wt. % Mo content (Figure 4c) are analyzed by EPMA point scanning. The results are shown in Table 2. The EPMA results show that the contents of Cr, B, and C in the eutectic structure are relatively high. Combined with XRD results, it can be concluded that the netlike eutectic structures are mainly composed of $M_2B$ and $M_7(C, B)_3$. The content of Fe in the dentrite is the highest, indicating that the dentrite structures are mainly composed of martensite and ferrite. From Figure 4 it can also be seen that the amount of martensite decreases, while the amount of ferrite gradually increases with the increase of Mo content. This is because Mo can promote the formation of ferrite [39]. Dendrite 2 contains a certain amount of Cr, which can strengthen the matrix by solid solution strengthening [40]. According to the EPMA and XRD results, the content of C in the granular eutectic structure is higher, and the granular eutectic structure is therefore identified as $M_{23}(C, B)_6$.

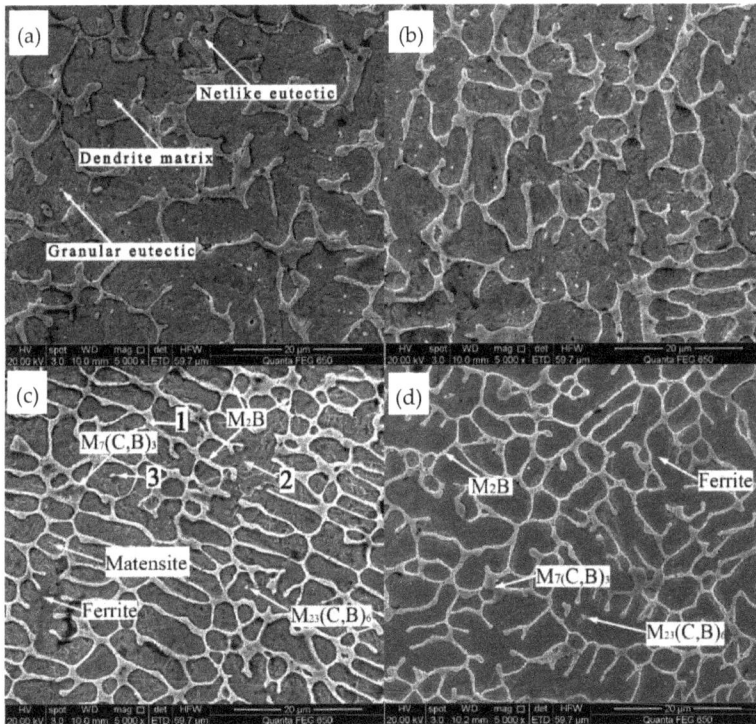

**Figure 4.** Microstructure of coatings with different Mo content: (**a**) Mo-free; (**b**) 2.0 wt. % Mo; (**c**) 4.0 wt. % Mo; (**d**) 6.0 wt. % Mo.

**Table 2.** The electron probe microanalyzer (EPMA) results of the point scan (wt. %).

| Point | Cr | C | B | Mo | Ni | Si | Fe |
|-------|--------|--------|-------|-------|-------|-------|--------|
| 1 | 19.978 | 4.877 | 4.789 | 4.674 | 1.910 | 0.428 | 63.344 |
| 2 | 15.336 | 2.606 | - | 1.801 | 1.937 | 0.783 | 77.537 |
| 3 | 14.224 | 11.034 | 0.139 | 1.780 | 1.697 | 0.779 | 70.347 |

Figure 5 presents the mapping scanning results of 4.0 wt. % Mo coating. It shows that the microstructures are inhomogeneous due to the severe elemental segregation that occurs during solidification. As can be seen from Figure 5b, e , f, the Ni, Fe, and Si elements are evenly distributed in the matrix phase. The Mo, Cr, and C elements (showed in Figure 5c,d,g, respectively) are rich at the grain boundaries, which indicates that the microstructures of the grain boundaries are primarily carbides of Mo and Cr. EDS component analyses were done in this area, and the relative contents (wt. %) of the composition calculated are as follows: 72.7% Fe, 15.2% Cr, 4.6% Mo, 5.2% C, 0.7% Si, and 1.6% Ni.

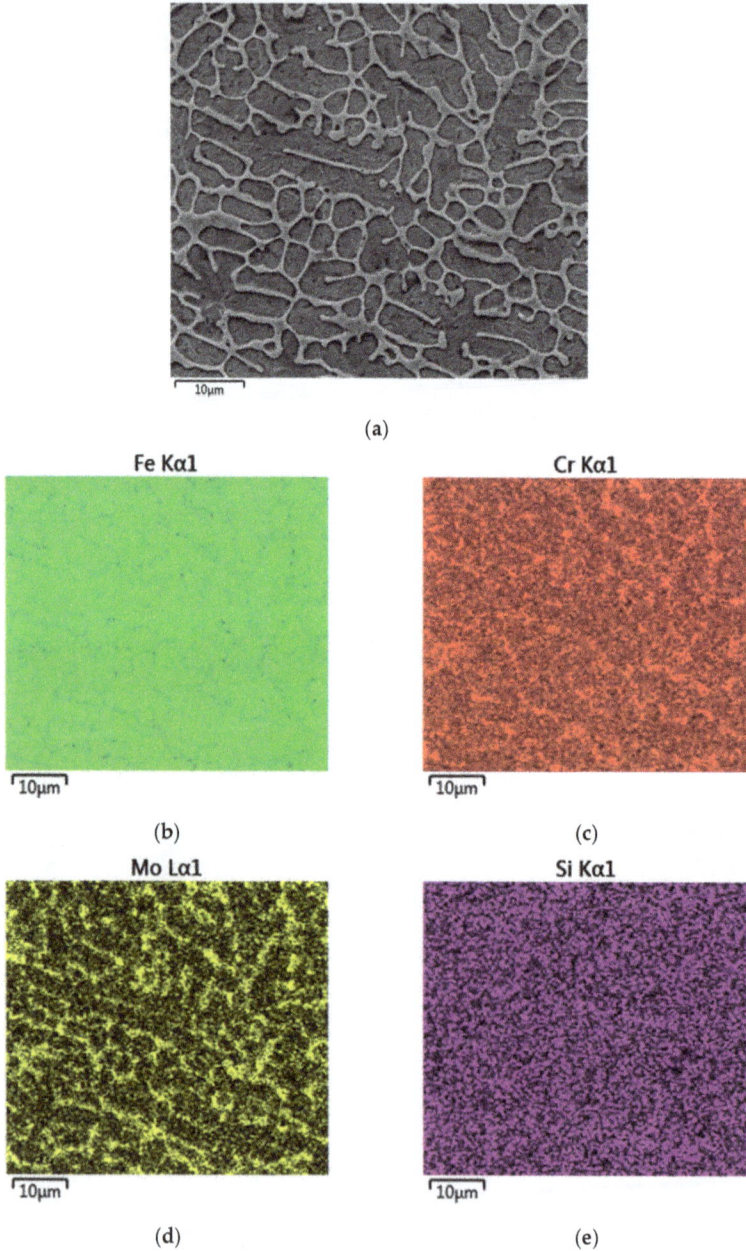

(a)

(b)

(c)

(d)

(e)

**Figure 5.** *Cont.*

Ni Kα1

C Kα1_2

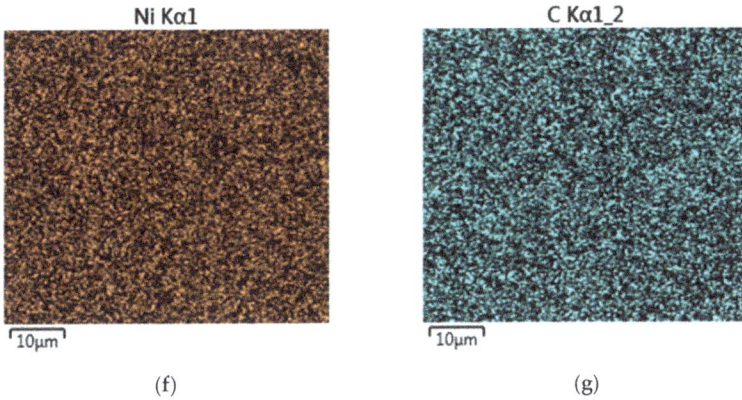

10μm

10μm

(f)

(g)

**Figure 5.** The mapping scanning results of the 4 wt. % Mo coating: (a) surface scan diagram (b) Fe; (c) Cr; (d) Mo; (e) Si; (f) Ni; (g) C.

## 3.2. Hardness Analysis

Figure 6 shows the morphology of the 4.0 wt. % Mo coating after the microhardness test. The size of the rhombus is basically the same, which indicates that the hardness distribution in the coating is uniform. The coating has good ductility because there are no obvious cracks around the indentations [41].

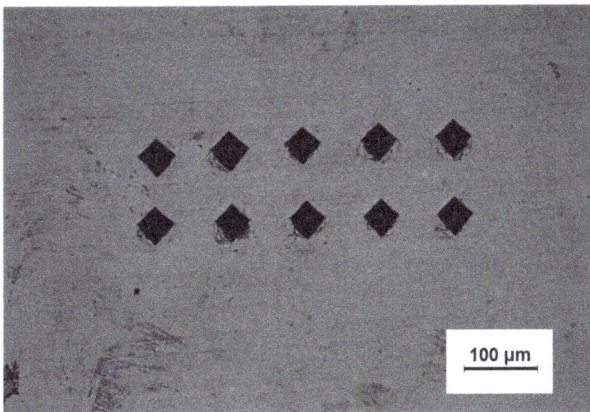

100 μm

**Figure 6.** Hardness indentation morphology of the coating (4.0 wt. % Mo).

The microhardness test results of the coatings with different Mo contents are shown in Figure 7. The microhardness decreases with the increase of Mo content in the coatings. The microhardness of the 6.0 wt. % Mo coating is 288.0 HV, which is about half of the microhardness of the Mo-free coating. The XRD results show that the amount of $M_{23}(C, B)_6$ increases and the contents of $M_7(C, B)_3$ decrease with the addition of the Mo element. The microstructure analyses show that the martensite decreases and the ferrite increases gradually with the increase of Mo content in the coating. It is known from the literature that the microhardness of martensite is 525.4 HV and the microhardness of ferrite is 235.1 HV [42]. The microhardness of $Cr_{23}C_6$ is 13.2 GPa and the microhardness of $Cr_7C_3$ is 18.3 GPa [43]. Therefore, the microhardness of the coating decreases gradually with the increase of Mo content.

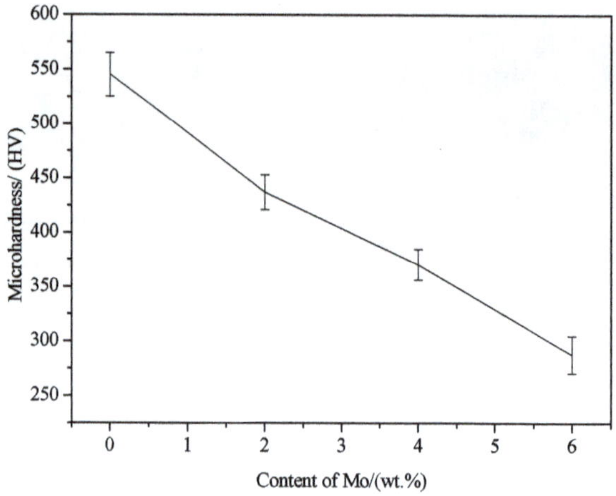

**Figure 7.** Effect of Mo content on the microhardness of the coating.

*3.3. Wear Resistance*

The wear loss and wear resistance of the coatings with different Mo contents are shown in Figure 8. The wear loss of the coatings decreases gradually with the increase of Mo content. The wear resistance of the 6.0 wt. % Mo coating is the best, which is about 3.7 times of that of Mo-free coating. Obviously, the addition of Mo in the stainless steel powder can greatly improve the wear resistance of the coatings.

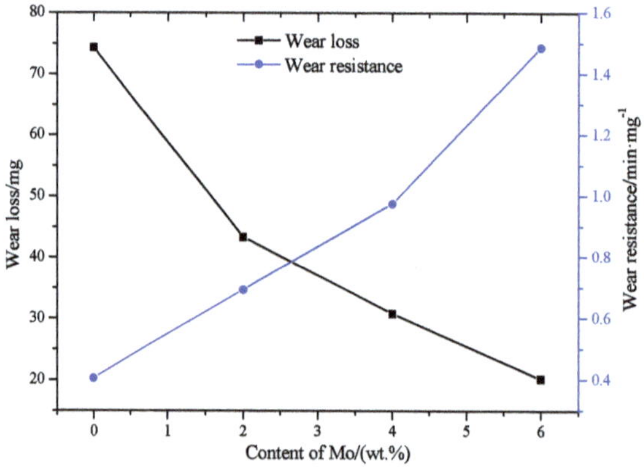

**Figure 8.** Effect of Mo content on the wear loss and wear resistance of the coatings.

To reveal the wear mechanism of the coatings with different Mo contents, the worn morphologies of the coatings are observed by SEM, as shown in Figure 9. The worn surface of the Mo-free coating (Figure 9a) has a large number of deep furrows, which are typical features caused by abrasive wear. When the contents of Mo are 2 wt. % and 4 wt. % (Figure 9b,c), the depths of the furrows are reduced and the amount of debris is gradually increased. Both abrasive wear and adhesive wear exist for these two cases. When the Mo content is increased to 6 wt. %, there is no furrow on the worn surface, but there is a large amount of debris. The main wear mechanism is adhesive wear. From the above analysis we can see that when the Mo content in the stainless steel coating increases, the wear mechanism of the coating changes from abrasive wear to adhesive wear.

In summary, the wear resistance of the stainless steel coating can be improved by adding the Mo element. The main reasons for the increase of wear resistance of the coatings are as follows. First, from the microstructure analysis it can be seen that the addition of Mo will refine the microstructure of the coatings, which can improve the wear resistance of the coating [44]. Second, the addition of Mo can change the phase types of the stainless steel coatings from the XRD results. $Mo_2C$ hard phase is formed when the Mo content is greater than 4 wt. %. Therefore, the wear resistance of the coatings can be effectively improved with the increase of Mo [45]. Third, a higher hardness generally represents a higher brittleness and lower ductility, and the lower hardness shown in Figure 6 therefore represents a ductility of the coating when Mo is added [46]. With the increase of Mo content, the hardness of the coating will decreases, and the ductility will increase. The increase of ductility will increase the wear resistance of the coatings [47]. Fourth, it is found from the worn morphologies that the depths of the furrows on the coating surfaces gradually decrease while the amounts of debris increase with the increase of Mo content. Such changes indicate that the wear mechanism is changed from abrasive wear to adhesive wear with the increase of Mo content. Therefore, the wear resistance of the coating is improved with the increase of Mo content.

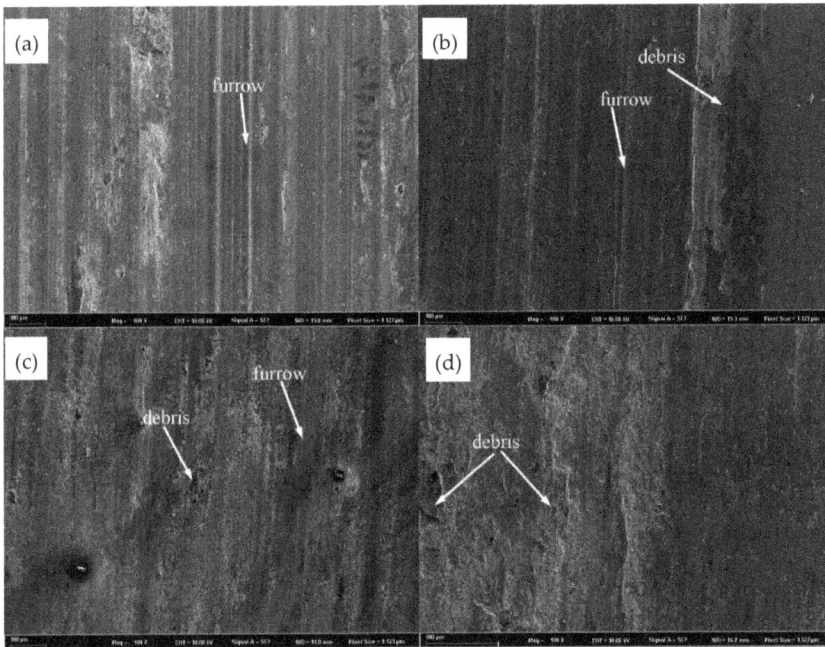

**Figure 9.** Worn surfaces of laser cladding coatings with different Mo contents: (**a**) Mo-free; (**b**) 2.0 wt. % Mo; (**c**) 4.0 wt. % Mo; (**d**) 6.0 wt. % Mo.

*3.4. Electrochemical Test Analysis*

The potentiodynamic polarization curves of the coatings with different Mo contents and 45 steel substrate are shown in Figure 10. The potential for all of the coatings is higher than that of the 45 steel substrate, which indicates that the corrosion resistance of the 45 steel substrate is significantly improved by the coatings. Laser cladding is a rapid heating and cooling process. From the XRD results, it can be seen that complex chemical reactions take place and new phases generate during the solidification process. The addition of Mo will refine the microstructure of the coatings, which can improve the corrosion resistance of the coating. With the increase of Mo content in the coatings, the potential first increases and then decreases. The coating with 2.0 wt. % Mo has the best corrosion resistance. The reasons are as follows. When the corrosion test is carried out in NaCl solution, the Cl⁻ will be adsorbed on the surface of the coating, which can react with the metal ion and form soluble chloride. This process can break the surface of the coating, and result in the formation of pitting (i.e., corrosion) on the surface of the coating. The addition of Mo can lead to a passivation phenomenon and greatly improve the pitting resistance of the coating [48]. When a certain amount of Mo is added in the coating, it can form a passivation film on the surface of the coating, which hinders the Cl⁻ entering the coating. So the pitting of the coating reduces and the corrosion resistance improves. When the Mo content is too high, the corrosion resistance decreases. Through the mapping scanning analysis, it can be seen that the Mo element is mainly enriched at the grain boundaries. Such an uneven distribution reduces the corrosion resistance of the coating.

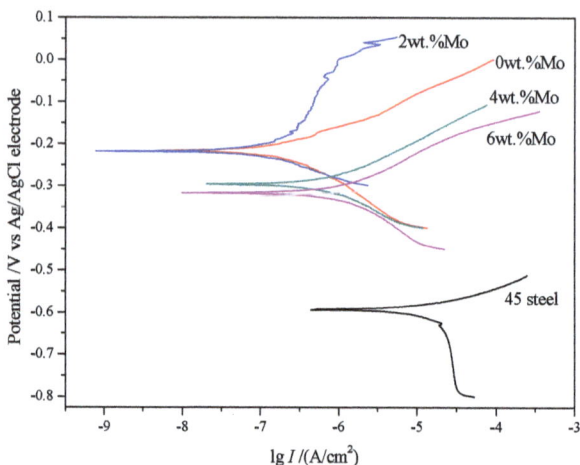

**Figure 10.** Potentiodynamic polarization curves of the substrate and four coatings with different Mo contents.

## 4. Conclusions

The stainless steel coatings with different Mo contents were prepared on the surface of a 45 steel substrate by laser cladding. The following conclusions can be obtained.

(1)  With the increase of Mo content, the amount of $M_7(C, B)_3$ decreases gradually, while the amount of $M_2B$, $M_{23}(C, B)_6$, and $Mo_2C$ gradually increases. When the Mo content is 4.0 wt. % and above, $Mo_2C$ phase appears in the coating.

(2)  The microstructure of the coatings is transformed from planar crystals to cell crystals, dendrites, and equiaxed crystals with the increase of distance from the fusion line. The amount of martensite decreases, while the amount of ferrite gradually increases with the increase of Mo content.

(3)  The microhardness decreases and the wear resistance of the coatings gradually increases with the increase of Mo content. The wear resistance of the 6.0 wt. % Mo coating is about 3.7 times that of the Mo-free coating.

(4)  The corrosion resistance of the 45 steel substrate is significantly improved by the laser cladding coating. With the increase of Mo content, the corrosion resistance of the coating first increases and then decreases. The coating with 2.0 wt. % Mo has the best corrosion resistance.

**Acknowledgments:** The authors appreciate the financial support for this work from National Natural Science Foundation of China (No. 51675303), the Open Foundation of Key Laboratory of E&M (Zhejiang University of Technology), Ministry of Education & Zhejiang Province (No. EM2016070103), and the Tribology Science Fund of the State Key Laboratory of Tribology (SKLT2014B09, SKLT2015B09).

**Author Contributions:** Kaiming Wang performed all experiments and wrote the paper; Baohua Chang and Hanguang Fu designed the experiments and reviewed the paper; Jiongshen Chen and Yinghua Lin performed the experiments; Yongping Lei analyzed the data.

**Conflicts of Interest:** The authors declare no conflict of interest.

## References

1. Vera, R.; Rosales, B.M.; Tapia, C. Effect of the exposure angle in the corrosion rate of plain carbon steel in a marine atmosphere. *Corros. Sci.* **2003**, *45*, 321–337. [CrossRef]
2. Storojeva, L.; Ponge, D.; Kaspar, R.; Raabe, D. Development of microstructure and texture of medium carbon steel during heavy warm deformation. *Acta. Mater.* **2004**, *52*, 2209–2220. [CrossRef]
3. Yang, Y.W.; Fu, H.G.; Lei, Y.P.; Wang, K.M.; Zhu, L.L.; Jiang, L. Phase Diagram Calculation and Analyze on Cast High-Boron High-Speed Steel. *J. Mater. Eng. Perform.* **2015**, *25*, 409–420. [CrossRef]
4. Kolman, D.G. A review of the potential environmentally assisted failure mechanisms of austenitic stainless steel storage containers housing stabilized radioactive compounds. *Corros. Sci.* **2001**, *43*, 99–125. [CrossRef]
5. Ronkainen, H.; Varjus, S.; Holmberg, K. Friction and wear properties in dry, water-and oil-lubricated DLC against alumina and DLC against steel contacts. *Wear* **1998**, *222*, 120–128. [CrossRef]
6. Gao, H.; Jiao, X.; Zhou, C.; Shen, Q.; Yu, Y. Study on Remote Control Underwater Welding Technology Applied in Nuclear Power Station. *Proc. Eng.* **2011**, *15*, 4988–4993. [CrossRef]
7. Birger, E.M.; Moskvitin, G.V.; Polyakov, A.N.; Arkhipov, V.E. Industrial laser cladding: Current state and future. *Weld. Int.* **2011**, *25*, 234–243. [CrossRef]
8. Simunovic, K.; Saric, T.; Simunovic, G. Different Approaches to the Investigation and Testing of the Ni-Based Self-Fluxing Alloy Coatings—A Review. Part 1: General Facts, Wear and Corrosion Investigations. *Tribol. Trans.* **2014**, *57*, 955–979. [CrossRef]
9. Han, B.; Li, M.; Wang, Y. Microstructure and Wear Resistance of Laser Clad Fe-Cr3C2 Composite Coating on 35CrMo Steel. *J. Mater. Eng. Perform.* **2013**, *22*, 3749–3754. [CrossRef]
10. Grum, J.; Žnidaršič, M. Microstructure, Microhardness, and Residual Stress Analysis of Laser Surface Cladding of Low-Carbon Steel. *Mater. Manuf. Process.* **2004**, *19*, 243–258. [CrossRef]
11. Zhao, W.; Zha, G.C.; Kong, F.X.; Wu, M.L.; Feng, X.; Gao, S.Y. Strengthening Effect of Incremental Shear Deformation on Ti Alloy Clad Plate with a Ni-Based Alloy Laser-Clad Layer. *J. Mater. Eng. Perform.* **2017**, *26*, 2411–2416. [CrossRef]
12. Paatsch, W. Energy turnaround—A challenge for surface technology. *Trans. Inst. Met. Finish.* **2016**, *94*, 228–230. [CrossRef]
13. Tan, H.; Luo, Z.; Li, Y.; Yan, F.; Duan, R.; Huang, Y. Effect of strengthening particles on the dry sliding wear behavior of Al2O3–M7C3/Fe metal matrix composite coatings produced by laser cladding. *Wear* **2015**, *324–325*, 36–44. [CrossRef]
14. Ray, A.; Arora, K.S.; Lester, S.; Shome, M. Laser cladding of continuous caster lateral rolls: Microstructure, wear and corrosion characterisation and on-field performance evaluation. *J. Mater. Process. Technol.* **2014**, *214*, 1566–1575. [CrossRef]
15. Peng, H.; Li, R.; Yuan, T.; Wu, H.; Yan, H. Producing nanostructured Co–Cr–W alloy surface layer by laser cladding and friction stir processing. *J. Mater. Res.* **2015**, *30*, 717–726. [CrossRef]
16. Xu, G.J.; Kutsuna, M. Cladding with Stellite 6 + WC using a YAG laser robot system. *Surf. Eng.* **2013**, *22*, 345–352. [CrossRef]
17. Gao, W.; Zhao, S.; Liu, F.; Wang, Y.; Zhou, C.; Lin, X. Effect of defocus manner on laser cladding of Fe-based alloy powder. *Surf. Coat. Technol.* **2014**, *248*, 54–62. [CrossRef]
18. Zeng, Q.; Sun, J.; Emori, W.; Jiang, S.L. Corrosion Behavior of Thermally Sprayed NiCrBSi Coating on 16MnR Low-Alloy Steel in KOH Solution. *J. Mater. Eng. Perform.* **2016**, *25*, 1773–1780. [CrossRef]
19. Zhou, S.; Dai, X.; Zheng, H. Microstructure and wear resistance of Fe-based WC coating by multi-track overlapping laser induction hybrid rapid cladding. *Opt. Laser. Technol.* **2012**, *44*, 190–197. [CrossRef]

20. Qu, S.; Wang, X.; Zhang, M.; Zou, Z. Microstructure and wear properties of Fe–TiC surface composite coating by laser cladding. *J. Mater. Sci.* **2008**, *43*, 1546–1551. [CrossRef]

21. Zhou, S.; Dai, X.; Xiong, Z.; Wu, C.; Zhang, T.; Zhang, Z. Influence of Al addition on microstructure and properties of Cu–Fe-based coatings by laser induction hybrid rapid cladding. *J. Mater. Res.* **2014**, *29*, 865–873. [CrossRef]

22. Wang, Y.; Zhao, S.; Gao, W.; Zhou, C.; Liu, F.; Lin, X. Microstructure and properties of laser cladding FeCrBSi composite powder coatings with higher Cr content. *J. Mater. Process. Technol.* **2014**, *214*, 899–905. [CrossRef]

23. Lei, Y.; Sun, R.; Lei, J.; Tang, Y.; Niu, W. A new theoretical model for high power laser clad TiC/NiCrBSiC composite coatings on Ti6Al4V alloys. *Opt. Lasers Eng.* **2010**, *48*, 899–905. [CrossRef]

24. Xu, J.S.; Zhang, X.C.; Xuan, F.Z.; Wang, Z.D.; Tu, S.T. Microstructure and Sliding Wear Resistance of Laser Cladded WC/Ni Composite Coatings with Different Contents of WC Particle. *J. Mater. Eng. Perform.* **2011**, *21*, 1904–1911. [CrossRef]

25. Qu, K.L.; Wang, X.H.; Wang, Z.K.; Niu, W.Y. Effect of Mo on the VC–VB particles reinforced Fe-based composite coatings. *Mater. Sci. Technol.* **2016**, *33*, 333–339. [CrossRef]

26. Ma, Q.S.; Li, Y.J.; Wang, J. Effects of Ti addition on microstructure homogenization and wear resistance of wide-band laser clad Ni60/WC composite coatings. *Int. J. Refract. Met. Hard. Mater.* **2017**, *64*, 225–233.

27. Lin, Y.; Lei, Y.; Fu, H.; Lin, J. Mechanical properties and toughening mechanism of TiB2/NiTi reinforced titanium matrix composite coating by laser cladding. *Mater. Des.* **2015**, *80*, 82–88. [CrossRef]

28. Zhang, G.Y.; Wang, C.L.; Gao, Y. Mechanism of Rare Earth CeO2 on the Ni-Based Laser Cladding Layer of 6063 Al Surface. *Rare Met. Mater. Eng.* **2016**, *45*, 1002–1006.

29. Weng, F.; Yu, H.; Chen, C.; Liu, J.; Zhao, L. Microstructures and properties of TiN reinforced Co-based composite coatings modified with Y2O3 by laser cladding on Ti–6Al–4V alloy. *J. Alloys Compd.* **2015**, *650*, 178–184. [CrossRef]

30. Hou, Q.Y.; He, Y.Z.; Zhang, Q.A.; Gao, J.S. Influence of molybdenum on the microstructure and wear resistance of nickel-based alloy coating obtained by plasma transferred arc process. *Mater. Des.* **2007**, *28*, 1982–1987. [CrossRef]

31. Wang, X.H.; Han, F.; Liu, X.M.; Qu, S.Y.; Zou, Z.D. Effect of molybdenum on the microstructure and wear resistance of Fe-based hardfacing coatings. *Mater. Sci. Eng. A* **2008**, *489*, 193–200. [CrossRef]

32. Ding, L.; Hu, S.; Quan, X.; Shen, J. Effect of Mo and nano-Nd2O3 on the microstructure and wear resistance of laser cladding Ni-based alloy coatings. *Appl. Phys. A Mater.* **2016**, *122*, 288. [CrossRef]

33. Wang, K.M.; Fu, H.G.; Lei, Y.P.; Yang, Y.W.; Li, Q.T.; Su, Z.Q. Microstructure and property of Ni60A/WC composite coating fabricated by fiber laser cladding. *Materialwiss. Werkstofftech.* **2015**, *46*, 1177–1184. [CrossRef]

34. Li, Q.; Lei, Y.; Fu, H. Growth mechanism, distribution characteristics and reinforcing behavior of (Ti, Nb)C particle in laser cladded Fe-based composite coating. *Appl. Surf. Sci.* **2014**, *316*, 610–616. [CrossRef]

35. Kurz, W.; Giovanola, B.; Trivedi, R. Theory of microstructural development during rapid solidification. *Acta. Metall.* **1986**, *34*, 823–830. [CrossRef]

36. Bansal, A.; Zafar, S.; Sharma, A.K. Microstructure and Abrasive Wear Performance of Ni-Wc Composite Microwave Clad. *J. Mater. Eng. Perform.* **2015**, *24*, 3708–3716. [CrossRef]

37. Liu, Z.; Qi, H. Effects of substrate crystallographic orientations on crystal growth and microstructure formation in laser powder deposition of nickel-based superalloy. *Acta Mater.* **2015**, *87*, 248–258. [CrossRef]

38. Li, B.; Jin, Z.; Ren, H.; Zhang, L.; Qu, W. Effect of molybdenum on microstructure and properties of low alloy high strength steel. *Heat Treat. Met.* **2016**, *41*, 145–147.

39. Shin, J.C.; Doh, J.M.; Yoon, J.K.; Lee, D.Y.; Kim, J.S. Effect of molybdenum on the microstructure and wear resistance of cobalt-base Stellite hardfacing alloys. *Surf. Coat. Technol.* **2003**, *166*, 117–126. [CrossRef]

40. Jarrett, R.N.; Tien, J.K. Effects of cobalt on structure, microchemistry and properties of a wrought nickel-base superalloy. *Metall. Trans. A* **1982**, *13*, 1021–1032. [CrossRef]

41. Li, B.; Jin, Y.; Yao, J.; Li, Z.; Zhang, Q. Solid-state fabrication of WC p-reinforced Stellite-6 composite coatings with supersonic laser deposition. *Surf. Coat. Technol.* **2017**, *321*, 386–396. [CrossRef]

42. Zhou, P.; Yang, L. Microstructure of 10CrMnMo dual-phase steel under different intercritical quenching temperature. *Heat Treat. Met.* **2017**, *42*, 110–114.

43. Min, T.; Gao, Y.; Li, Y.; Yang, Y.; Li, R.; Xie, X. First-Principles Calculations Study on the Electronic Structures, Hardness and Debye Temperatures of Chromium Carbides. *Rare Met. Mater. Eng.* **2012**, *41*, 271–275.

*Appl. Sci.* **2017**, *7*, 1065

44. Hou, Q.Y.; He, Y.Z.; Gao, J.S. Microstructure and wear resistance of Mo /Ni-based alloy coating produced by plasma cladding. *Chin. J. Nonferrous Met.* **2006**, *16*, 1595–1602.

45. Xiong, Z.; Xu, Q.; He, Q.; Wang, Y. The effect of Mo content on the organization and the performance of the hypoeutectic high carbon, high chromium alloy. *Electr. Weld. Mach.* **2016**, *46*, 59–62.

46. Liu, Z.J.; Liu, C.; Chen, H.; Li, Y.K.; Cheng, J.B.; Su, Y.H.; Liu, D. Impact wear-resistant hardfacing austenitic material. *Trans. China Weld. Inst.* **2005**, *26*, 9–12.

47. Hu, Y.W.; Yin, Y.S. Analysis of affecting factors of wearing resistance of surfacing layer. *J. Shenyang Univ. Technol.* **2002**, *24*, 386–388.

48. Li, D.; Zhou, F.; Yu, S. Microstructure and corrosion resistance of FeCrNiMnMoxB0.5 high entropy alloy coating prepared by laser cladding. *High Power Laser Part. Beams* **2016**, *28*, 1–6.

MDPI AG

St. Alban-Anlage 66

4052 Basel, Switzerland

Tel. +41 61 683 77 34

Fax +41 61 302 89 18

http://www.mdpi.com

*Applied Sciences* Editorial Office

E-mail: applsci@mdpi.com

http://www.mdpi.com/journal/applsci